Early Aviation

in

Monmouth County
New Jersey

George Joynson

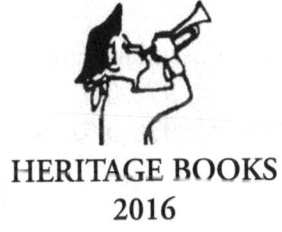

HERITAGE BOOKS
2016

HERITAGE BOOKS
AN IMPRINT OF HERITAGE BOOKS, INC.

Books, CDs, and more—Worldwide

For our listing of thousands of titles see our website at
www.HeritageBooks.com

Published 2016 by
HERITAGE BOOKS, INC.
Publishing Division
5810 Ruatan Street
Berwyn Heights, Md. 20740

Copyright © 2016 George Joynson

All rights reserved. No part of this book may be reproduced or transmitted in any form or by any means, electronic or mechanical, including photocopying, recording or by any information storage and retrieval system without written permission from the author, except for the inclusion of brief quotations in a review.

International Standard Book Numbers
Paperbound: 978-0-7884-5701-2
Clothbound: 978-0-7884-6370-9

Dedication

Early Aviation in Monmouth County is dedicated to my dad, with love. Happily retired at Venice on the Isle, he always taught me to do the right thing, and sometimes I listened.

Contents

1. Charles J. Hendrickson — 1
2. America's Greatest Aviation Meet — 11
3. Monmoth Aviation Patents — 39
   - Thomas Walling — 39
   - Clarence Moore — 49
   - William Hurst — 54
   - Stephen Stevens — 59
   - J. Thomas Havens — 62
4. Gala at Rest Hill — 65
5. Aeromarine Plane and Motor Company — 81
6. Casey Brothers — 99
7. De Luxe Air Services, Inc. — 117
8. Monmouth Builders and Flyers — 125
   - Ralph Bray — 125
   - S. Willis Wood — 129
   - Edward Meyer — 130
   - M. Frank Morris — 135
   - Peter Kimmerland — 136
   - Leon Morford — 136
   - Howard Borden — 137
   - Patrick Byrne — 137
   - Edward Schroeder — 138
- Sources — 139
- Index — 149

Acknowledgments

I'd especially like to thank David Bohny, whose knowledge and assistance in technical and computer issues was greatly appreciated. Also, thank you to Linda Bohny and Barbara Marshall, who were a huge help with editing the text. Thank you to the aviators' family descendants who shared their family photos and entrusted me to tell the story of their ancestors, including Kathleen Bartlett, Reverend William Breslin, Ardy Tobin, Jolie Morris, Kris Davis, Debbie Nelson, Dick Winters, Sally O'Brien, Dorothy Herbert Yeargin, Nancy Philip Byrne and Colin Hunt.

Thank you to the librarians, authors, archivists, museum personnel and collectors who helped, including Shea Oakley, Daniel Kusrow and Michelle Fanelli of the New Jersey Aviation Hall of Fame; Peggy Dellinger and Virginia Richmond of Township of Ocean Museum; Jack and Angel Jeandron of Keyport Historical Society; Robert W. Stewart and Kathleen Meigar of Asbury Park Library; Sister Debbie Drago of Sisters of the Good Shepherd and Jeanne Navagh of Collier Services; Maury York of Empire State Aerosciences Museum; Elizabeth McDermott and Dawn Bladzinski of the Red Bank Library; Chuck Watson of Morgan Stanley, George and Kathy Severini of Dorn's Classic Images; and Roberta Van Anda.

I am very appreciative of the institutions that were willing to grant permission to use their photographs, including Asbury Park Library, Red Bank Library, New Jersey Aviation Hall of Fame, Empire State Aerosciences Museum, Smithsonian Institute Air and Science Museum, Library of Congress, National Archives, Wright State University and Missouri State Archives. Thank you to anyone who helped me with this project who I accidently forgot to include in the above list.

Introduction

This book is more about people and events in Monmouth County than it is about the history of aviation. It is a tribute to the brave souls who dared try to fly. Readers will learn about local residents who designed, built, patented, flew, or crashed in, early flying machines. Others worked at the Aeromarine aeroplane manufacturing plant in Keyport. Locals had the opportunity to attend the 1910 Aviation Meet in Asbury Park, and Robert Collier's 1911 Aviation Gala in Marlboro. Some had the chance to meet visiting designers and aviators, including aeroplane patentee Thomas Edison, Wilbur Wright and his five-man Wright Exhibition Team, and stunt flying daredevils like Ruth Law and her Flying Circus Act.

Worldwide interest in human flight exploded on December 17, 1903, when Wilbur and Orville Wright proved that successful manned flight was possible. Wright's machine-powered aeroplane is seen as one of the greatest inventions of all time. Their historic fifty-nine second flight set off a frenzy, locally and worldwide, of aspiring young men who dreamed of building a flying machine of their own design and making a flight in it.

Just like Kitty Hawk, Monmouth County was an ideal location for experimental flight, with slight hills and coastal winds for gliders, and flat terrain and plowed fields for landing machine powered flights.

These early aviators risked their lives, their reputations, and their worldly possessions to pursue their dreams of flight. The 1910s were crazed with men and women from Asbury Park, Long Branch, and Red Bank who dreamed of flying. Some just wanted to see if they could do it. Others wanted desperately to fly higher, faster or farther than the previously reported flight. Records were broken sometimes weekly during these early years. Inventors confidently thought their design was the breakthrough that would solve some problem. Some flights were successful; some were deadly. Regardless of the risk, "aviator fever" was rampant in Monmouth County and it blanketed the nation as well. Aero clubs sprung up; businessmen organized aviation exhibitions; contests were held and prizes were won.

Promoters offered rewards to spur them on, as if the personal reward of flight wasn't enough. Locally, James A. Bradley, Senator, Asbury Park mayor, editor, and real estate developer, offered a $500 award to the owner, lessee or manipulator of the first airship to use the Asbury Park Athletic Ground as the point of destination or landing on a trip from New York. Bradley made the offer July 4, 1908, and said his offer was good for 1½ years. On a grander scale, in 1910, Randolph Hearst offered a $50,000 reward to the first aviator to fly from coast-to-coast in less than thirty days.

In 1910, a spokesman for the US Patent Office in Washington, DC, estimated that applications for aeroplane designs were flooding into the office at the rate of about 150 per week! A large majority of them were rejected and those records are not available. These men and women had one thing in common, that they wanted urgently to fly.

The press called them "birdmen." They also called them aviators, aeronauts, daredevils, and barnstormers. They flew in things called a flying machine, airship, aeroplane, monoplane, biplane, or hydroaeroplane. Usage of the English spelling "aeroplane" decreased around WWI as the American version "airplane" became more common. Back then there were great debates. Was the biplane, monoplane, or triplane the safest or more successful? How about a plane with five wings? Was one, two, or more propellers better for stability?

The dawning of human flight brought out American ingenuity, determination, and persistence. It attracted men who were not afraid to fail, the inspired, curious, optimistic, and patient experimenters, who had strong faith that their idea would revolutionize science and at the same time make them wealthy and famous. Despite the experiments, the patents, patent infringement lawsuits and countersuits, men were willing to try anything to fly like a bird. Some sunk their life savings into their project. New records came and went for the highest, fastest or longest flight achieved.

We've come a long way since Wright's first flight. It's doubtful that they could have imagined today's accomplishments. In a little more than one hundred years, we have flown faster than 2,000 mph, and higher than 367,000 feet. We have broken the sound barrier, achieved solar powered flight around the world, and a manned trip to the moon. Recently a neighbor demonstrated a remote controlled, tri-helicopter drone equipped with high-definition video cameras. I can't imagine what the next hundred years will bring.

Some topics, such as the Red Bank Airport, Asbury Park Meet, and Aeromarine Plane & Motor Company, deserve more than a chapter. I invite you to learn about aviation in Monmouth County during the early years of 1908-1930. Want to know who was building, designing, and flying? What worked and what didn't? Please read on.

1. Charles J. Hendrickson (1879-1948)

Charles John Hendrickson,
first known Monmouth County resident to fly, March 29, 1909.
Courtesy of Katharine Bartlett.

Charles John Hendrickson, aka CJ, was the earliest known resident of Monmouth County to fly. Hendrickson was born on August 7, 1879, in Middletown, New Jersey, the son of John S. Hendrickson. Hendrickson was born into a family that practiced farming for many generations but he had other ideas and did not follow his family's tradition. About 1904, Hendrickson spent two years in the Mechanical Engineering program at Lehigh University then two more years at Harvard. By1910, CJ was a thirty-year old unmarried merchant living with his widowed father on Kings Highway in Middletown. A year later he married Adele G. Taylor and began working for Western Electric Company.

Hendrickson was an inventor and an aeroplane builder but did not patent any of his flying machines. He did, however, invent an automatic switching device which he patented in 1916 and assigned his patent rights to Western Electric. In 1917, he sold his father's five-acre farm, with terra cotta house and barn in Middletown, and moved to New York City, where he was living when the US Army called him into service on March 25, 1918.

Middletown has a Biplane Inventor

During the winter of 1907-1908, Hendrickson was busy building an aeroplane in his father's barn. A few of his close friends knew what he was doing but Hendrickson tried to keep his work secret. A unique design of an aircraft in the early development of flight, if successful, could be enormously valuable. Hendrickson was protecting his ideas from being copied or stolen.

Hendrickson's biplane, photo by Edwin Levick.
Smithsonian Institute, National Air & Space Museum (NASM 81-15606).

The biplane he built was thirty-three feet wide and nine feet long. He used light wood and metal for the framework and wide stretches of canvas to cover the top and bottom wings. He rigged the pilot's seat in the center of the machine and crafted a seven-foot propeller out of oak. Hendrickson confidently believed his flying machine would be a practical craft for aerial navigation.

Biplane Trials

In February 1908, Hendrickson tested his new machine. His biplane rose in the air for a few feet, and then turned upside down. In his early attempts, Hendrickson obviously grappled with the issue of aeroplane stability. On that first flight test, his plane was damaged but Hendrickson was not hurt. A month later, with repairs completed, he scheduled a second test on April 18. That Saturday morning there was little wind blowing and his aeroplane engine had not yet been installed, but he and his helpers worked to properly balance the machine.

Stability Issue

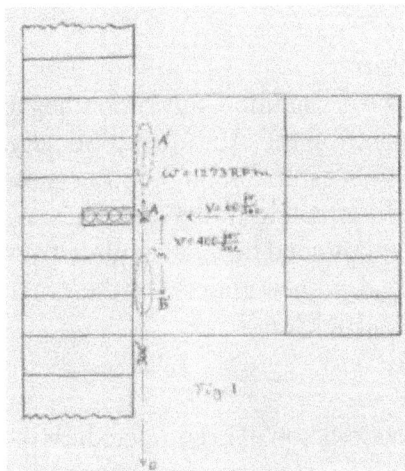

Hendrickson wrote "The Stability of Aeroplanes," published in the August 1908 edition of *Aeronautics, The Magazine of Aerial Navigation.* Hendrickson wrote that "a single propeller creates unbalanced forces acting to destroy equilibrium and that these forces are great enough to warrant consideration" [of using two propellers]. In the early days of aviation, the number of propellers was a great debate among those afflicted with "aviation fever." Hendrickson's drawing from "The Stability of Aeroplanes." *Aeronautics, August 1908.*

A year later, Hendrickson announced he was building a new airship of his own design and claimed it would outstrip all rivals in aerial navigation. This plane had large wings in the shape of a bird. He rigged a jib sail in the rear to steer it, and added a canvas seat for the operator. It had a shaft, motor, two propellers and two batteries.

He tested it March 27, 1909. Hendrickson started his trial by positioning his machine at the top of a small knoll. He would run with it down the hill then hop into the seat. Despite not much wind that Saturday morning, Hendrickson flew in his new machine.

Witnesses reported Hendrickson's plane carried him about fifteen yards. Those fifteen yards must have been a thrilling adventure - his personal reward for his persistence, ingenuity and determination. With the success in his second machine, Hendrickson joined the newly formed Aeronautical Society of New York.

In June 1909, Hendrickson agreed to exhibit his monoplane glider at the Aeronautic Society Exhibition at Morris Park in New York. He disassembled his aeroplane and had it shipped "knocked down" by express, from Middletown, New Jersey, to the Aeronautics Grounds at the Morris Race Track in Westchester County, New York. Express shipping companies did not yet have a set rate for shipping aeroplanes, so they charged him a mere $1.25.

Morris Park Aeronautic Convention

On June 26, 1909, the Morris Park Aeronautic Convention began its first exhibition of flying machines for the Aeronautic Society. This event was a huge success. Crowds estimated at 5,000 or more came out to see men in flight and their amazing flying machines. As part of the exhibition, Hendrickson announced his plan to glide from the grandstand roof in his own newly designed glider. For this stunt, the press dubbed him "Daredevil Hendrickson."

Chanute Cup

At the Morris Park Convention Hendrickson also competed for the Chanute Cup in Gliding off a Mound Category. Also competing were three other gliders: Charles R. Wittemann with his brother Adolph Wittemann who were starting to develop flying machines in Staten Island, and William H. Aitken of Chester, Pennsylvania. The Wittemann brothers were experienced glider manufacturers.

Aitken was also an experienced glider and author of "How to Glide," published in *Aeronautics*, June 1909. Aitken wrote that all you need to glide was a well-built glider, about a twenty-foot hill, two assistants, wind, and lots of courage. Aitken had recently demonstrated his gliding skills at the Philadelphia Country Club. His glider had wings measuring 20 feet by 4½ feet. After the Morris Park Meet, Aitken headed to Great Neck Bluffs on Long Island, New York, for more demonstrations.

CJ Hendrickson's glider; photograph, drawing and applied paint.
Smithsonian Institute, National Air & Space Museum (NASM 00052254).

Wittemann glider in towed flight in 1908.
Courtesy of NJ Aviation Hall of Fame.

Octave Chanute was there to watch the competition for the cup he sponsored. Chanute said the grounds at Morris Park were "ideal for aeroplane flights." Chanute was fascinated with aviation as early as the 1890s when he wrote "Progress in Flying Machines," published in 1894 by the *American Engineer and Railroad Journal*, New York. Chanute's article contained many pictures of some very early designs of flying machines, mostly resembling bird shapes. He wrote that aeroplanes should be capable of holding at least three passengers.

Octave Chanute sponsored "Gliding off a Mound" contest at the Morris Park Aviation Meet in 1909. *Library of Congress.*

Demoiselle Glider

Hendrickson's glider may have had some influence on the design and construction of Alberto Santos-Dumont's aeroplane, *"Demoiselle." A*ccording to the 1909 *Aeronautic Society's Bulletin No. 1*, "The second apparatus was a strikingly interesting glider designed by CJ Hendrickson, which foreshadowed to a singular degree now famous '*Demoiselle'* by Santos-Dumont." Alberto Santos-Dumont was from Brazil but built experimental gliders in France. He used bamboo to construct the framework of his lightweight monoplanes. Santos-Dumont wanted to advance aviation. Instead of patenting, Santos-Dumont gave *Popular Mechanics* his drawings and instructions on how to build the *Demoiselle*, which they published in the June 1909 edition.

This is how Santos-Dumont got his *Demoiselle* to the airfield. *Library of Congress.*

Alberto Santos-Dumont.
Library of Congress.

The *Demoiselle* in flight in 1909.
Courtesy of NJ Aviation Hall of Fame

Hendrickson represented the C.E. Conover Co. at the Morris Park Aviation Meet. The C.E. Conover Company manufactured aero cloths and chemicals like Naiad Aero Varnish to treat any material. The company advertised Naiad as safe, "non-inflammable (sic)," tight, strong, smooth, weather proof, tough, flexible and transparent. Headquarters were located at 101 Franklin Street in New York City.

Aeronautics, October 1912.

WWI Military Aviation

Hendrickson's experience and enthusiasm as a designer and builder of experimental aeroplanes paved the way for his training and service during WWI. At age thirty-eight, Hendrickson registered on December 1, 1917, in the Reserve Corps at New York and was called into service on March 25, 1918.

Ground School

At the start of WWI, the US Army lacked trained pilots. It established an Aviation Cadet Training Program to fill the void. The School of Military Aeronautics at Ohio State University in Columbus, Ohio, was a twenty-week course of instruction with ninety hours of flying time. Known as "ground school," the program was only open to men who had completed two or more years of college. The cadets studied aerodynamics, aeroplane flight, rigging and repairs, map reading, internal combustion engine, combat, and machine gun operations.

Hendrickson completed the Military Aeronautics training course at Ohio State on January 14, 1918. From Ohio, Hendrickson went to the School of Military Aeronautics at Cambridge, Massachusetts, where he remained until honorably discharged on March 24, 1918. The next day, Hendrickson accepted a commission as Second Lieutenant with the Sixth Aero Squadron and shipped to Ford Island on Honolulu, Oahu, in the Territory of Hawaii.

Sixth Aero Squadron

The Sixth Aero Squadron was organized December 1916 with the purpose of establishing a seaplane base in Hawaii. It was officially activated March 17, 1917, when the first forty-nine men arrived under Captain John B. Brooks. Luke Field on Ford Island, as it was later named, was equipped with a Curtiss JN4-D and two Curtiss N-9 seaplanes.

The JN4-D was a biplane with a wingspan of forty-four feet, top speed of 75 miles per hour, and service ceiling of 6,500 feet. It had two inline seats, designed for instructor and pilot in training. The military used it extensively during WWI. This *Jenny* model was introduced in 1917 by the Curtiss Aeroplane Company at Hammondsport, New York. Curtiss built more than 6,000 of them.

Curtiss JN4-D, commonly known as the *Jenny*.
www.wickapedia.com.

The Curtiss N-9 was a hydroplane, fitted with a single central pontoon. It was slightly bigger than the JN4s, with a fifty-three-foot wingspan, maximum speed of 75 miles per hour, and a service ceiling of 7,000 feet.

Hendrickson was stationed at Ford Island until he received an honorable discharge on May 27, 1919. After completing his military service, Hendrickson returned to New York City and worked as an engineer for the Bell Telephone Company.

He retired after thirty-one years of service at Bell Laboratories in February 1946. Charles Hendrickson died January 19, 1948. The home-made oak propeller that Hendrickson built in the winter of 1907-1908 is now in possession of his granddaughter.

Hendrickson retired in 1946. *Courtesy of Katharine Bartlett.*

2. America's Greatest Aviation Meet

Wilbur Wright came to the Asbury Park Aviation Meet in 1910.
Library of Congress.

"America's Greatest Aviation Meet," held August 10-27, 1910, was the first aviation meet ever held in Monmouth County, New

Jersey. The seventeen-day event attracted many visitors and heavy press coverage. The event followed other successful aviation meets, in Atlantic City in July, and earlier that year in Los Angeles and Indianapolis.

From *left*, Frank Coffyn, Augustus Roy Knabenshue, and Walter Brookins. Knabenshue negotiated the contract for Wright Brothers Exhibition Team to attend the Asbury Park Meet. *Courtesy of Empire State Aerosciences Museum.*

Representing the Wright Brothers Company, agent Augustus Roy Knabenshue signed a contract with the Aero and Motor Club of

Asbury Park and Asbury Park Business Men's Association, to supply five experienced pilots and aeroplanes. Knabenshue was an aviator who set several world records in dirigibles. In March 1910, he presented his idea to the Wright brothers of assembling a team of well-trained pilots and booking them for exhibitions. The Wright Brothers Company and the Aero Club agreed to let Colonel Mahlon R. Margerum plan and direct the events. Margerum had successfully organized the New Jersey State Fair.

Colonel Mahlon R. Margerum directed the New Jersey State Fair and the Asbury Park Aviation Meet. *Genealogical and Personal Memorial of Mercer County, New Jersey, Francis Bazley Lee.*

Margerum set up committees for advertising, entertainment, finance, grounds, tickets, parking, music, publicity, grandstands, concessions, prizes, medicinal and police. He hired Cole and Company of 401 Cookman Avenue in Asbury Park to be official photographers for the event. John McCaffrey of United States Tent and Awning Company of Chicago, Illinois, was hired to construct a canvas wall outlining the field, and tents for the aviators, press and medics.

The Aero Club published an Official Souvenir Program. Local business owners advertised in the program, which helped defray the event expenses. Margerum set a schedule of flights everyday starting at 4:00 P.M. Events also included kite flying contests, hot air balloons, dirigible ascents and amateur competitions.

Terms of the contract specified that the first $20,000 in gate receipts went to the Wright Brothers Company; the second $20,000 would go to the shareholders of the Aero and Motor Club of Asbury Park. Any profit after that would be split fifty-fifty. Aviators would receive twenty dollars per week and fifty dollars per flight. A seat in the grandstand cost fifty cents; an automobile parking spot cost two dollars.

The Asbury Park Air Meet was held in Interlaken, a neighborhood in Ocean Township, separated from Asbury Park only by Deal Lake. *Postcard Courtesy of Asbury Park Library.*

On a forty-acre field in Interlaken, north of Deal Lake in Ocean Township, they built a grandstand that held 15,000 spectators. They set up tents along one side of the field and dedicated space for parking 3,000 cars. Out of town visitors could reach the event by train, just a short walk from the North Asbury Park train station. They set up fifty tables with Monarch typewriters in the press tent. The Aero Club asked Company H of the New Jersey National Guard to patrol the grounds, which they did in uniform and on horseback. Margerum dubbed it America's Greatest Aviation Meet.

The Wright Brothers Company cleared their $20,000 by receipts taken in at the end of the eighth day. Shareholders of the Aero Club were not as lucky. By the eighth day, their cut was only $5,900 before expenses for prize money, building a grandstand, tent rentals, etc.

Wilbur Wright did not fly, he said, because of "being involved in too many law suits." Orville did not attend the meet. The contract did not call for either Wright to appear personally but Wilbur did travel from New York City to come out to observe some flights and visit his injured Wright Exhibition Team aviator, Walter Brookins.

Colonel Margerum scheduled the ten-day event to start August 10. Due to bad weather causing cancellation of two days of flights, and the desire to increase gate receipts, flights continued for an extra seven days, ending on August 27. Estimates ran as high as 75,000 guests the first day. Special days included: Children's Day, Governor's Day, Aero Club Banquet Day, Scientist Day and Amateur Day. On Children's Day, all children under the age of twelve were admitted free if accompanied by one paid adult, but all flights scheduled on Children's Day were cancelled due to high winds and rain.

Aero and Motor Club of Asbury Park

Local real estate broker Milan Ross was a shareholder of the Aero Club. *The Jersey Coast in Three Centuries, 1902.*

The Aero and Motor Club of Asbury Park had incorporated just a few weeks earlier, on July 16, 1910. Thirteen men and one woman received ten shares each: three hotel proprietors Charles A. Atkins, Mrs. Margaret H. Frost, and Alonzo R. Parsons; Dr. Thomas H. Pratt; civil engineer John W. Aymar; bank president James M. Ralston, and bank cashier William A. Berry; publicity bureau superintendent Harold E. Denegar, Sr.; newspaper publisher J. Lyle Kinmonth; real estate brokers George W. Pittenger, and Milan Ross; stockbroker Harry J. Rockafeller; retired James G. Warner and garage proprietor Charles R. Zacharias. Pittenger was president.

Mrs. Margaret Frost was a widowed proprietor of the Hotel Lafayette on Fourth Avenue in Asbury Park, where the Wright Exhibition Team aviators stayed during the event. Wilbur Wright and Walter Brookins stayed as guests of Mrs. Abercrombie Fell in Bayhead.

Wright Exhibition Team

Wright agreed to have five aviators appear at this event: Frank T. Coffyn, senior aviator, Walter R. Brookins, Archibald Hoxsey, Ralph Johnstone, and Duval La Chapelle.

Wright had trained these five flyers and called them his pupils. He called them the most daring aviators in the world. Brookins and Coffyn came to the meet as current world aviation record holders. In 1910, aviation records for altitude, speed, turns, duration and other categories, were being broken frequently. It was also a fatal year. Also in 1910, fifty-three aviators lost their lives in aviation advancement or pursuit of prize money, including Hoxsey and Johnstone. Aviators at the Asbury Park Aviation Meet also included hot air balloonists Fred Owens and Johnny Mack.

Frank T. Coffyn (1878-1960)

Frank Trenholm Coffyn was born in Charleston, South Carolina. Wright trained him in April and by the time of the Asbury Park Meet, designated Coffyn as the head aviator. Coffyn was considered the sensible flyer. He wanted the flights cancelled in heavy weather, but the others were determined that the show must go on. He later appeared in several silent movies.

Frank T. Coffyn. *Courtesy of Special Collections and Archives, Wright State University.*

Archibald Hoxsey. *Library of Congress.*

Archibald Hoxsey (1884-1910)

Archibald Hoxsey, Arch, or "Hox" as he was sometimes called, was born October 15, 1884, in Staunton, Illinois. He worked as a skilled mechanic in the Wright brothers' aeroplane factory before learning to fly. On December 26, 1910, Hoxsey flew to an altitude of 11,474 feet, setting the world flight altitude record. He died five days later trying to break his own record.

Walter Brookins. *Courtesy of Empire State Aerosciences Museum.*

Walter R. Brookins (1888-1953)

Walter Richard Brookins grew up in Dayton, Ohio, as a neighbor of the Wright family. At the Asbury Park Meet he was just twenty-two years old. Brookins came to Asbury Park owning the world record for highest flight at a measured altitude of 6,275 feet, set a few weeks earlier on July 9, in Atlantic City.

Ralph Johnstone. *Courtesy of Empire State Aerosciences Museum.*

Ralph Johnstone (1880-1910)

Ralph Johnstone was born on September 18, 1880, in Parsons, Kansas. Johnstone told his father, "When I make a flight, I have my plans well laid. Before I leave the ground, I know exactly what I am going to do. Don't worry about me being injured." Knabenshue hired Johnstone as part of the Wright Exhibition team to do low altitude roller coaster rolls at fifty to one hundred feet for the Asbury Park Meet.

Wilbur Wright said of Johnstone, "If I leave that boy alone for a minute, he'd be in the hospital." Johnstone was an aviator trickster. When Brookins crashed and was not able to fly for a few days, Johnstone wanted to take his assignments. He wanted to perform a stunt called loop-the-loop, and tried to convince Wright that he could do it. Johnstone said:

> "I've got it all figured out. All I have to do is go up, say 500 feet, point her down at the sharpest angle she'll stand, and then at just the right moment, lift the elevating planes. She can't go wrong. I tell you, she can't go wrong."

Two months later Johnstone died in a fatal crash.

Duval La Chapelle (1869-1932).

Rare photo of French aviator Duval La Chapelle, *center,* between Coffyn, *left,* and Brookins, *right. Courtesy of Empire State Aerosciences Museum.*

Duval La Chapelle, the "French sky pilot," did not fly as part of Asbury Park Meet. Chapelle completed his training at the Wright Company in Dayton, Ohio, in June 1910. He competed at the Indianapolis meet and held the aviation speed record in Canada.

Opening Day

Aviator Walter Brookins was the first to go airborne on Opening Day, August 10, at Asbury Park. His flight started off smooth as he soared to 300 feet then circled the landing field three times. In attempting to land, Brookins crashed into the crowd, injuring a dozen spectators. He broke five teeth and suffered a broken nose. His Wright biplane was a total loss, estimated at $7,500 dollars.

As he approached the landing field, a crowd of newspaper photographers had gathered, forcing him to abort the landing. He tried to re-ascend but did not have enough speed. Brookins steered his craft into an alley, avoiding the filled grandstand. Several police officers and young boys selling programs were standing in the alley. Brookins crashed from about twenty-five feet up, as the nose of the plane hit first, digging into the ground. Before they took him off the field, he told Margerum, "I did the best I could colonel, I did not hit the grandstands."

While medics attended the injured, Arch Hoxsey prepared his plane to get airborne. Within forty-five minutes of Brookins' crash, Hoxsey was up in the air for the second exhibition flight. The crowd cheered as he executed a perfect landing. Brookins' brave effort generated much press coverage, which in turn may have attracted more paying visitors to the meet.

Brookins went to the Fell residence at Bay Head to recover. For a few days he was sore and stiff, with a puffed face and bandaged nose. When Coffyn came back from visiting him, he said Brookins was full of grit and determined as ever to get back up in the air. Brookins told Coffyn he planned to fly the next Monday.

Wilbur Wright arrived at Asbury Park the next day to watch the flights and investigate Brookins' crash landing. He traveled to Bay Head to visit Brookins and hear the story straight from him. Wright ordered a replacement plane, which luckily was ready to be shipped from Dayton, Ohio.

From then on, Wilbur Wright ordered his aviators to not fly directly over the field or the grandstand. Regarding Brookins' first day crash, Wilbur Wright said,

> "The youngster was over ambitious and he had to learn what an older man would have known offhand, never go up from the field while the people were passing over it. I had to learn that in my early efforts in France. Every time I got ready to fly and the field was not clear, I simply wheeled my machine back into the shed and waited. Pretty soon they learned they had to keep off the field or see no flying and they learned how to act."

Injured Spectators

Spectator Maurice Gorsuch of Asbury Park suffered a broken arm when Brookins crashed. Representing Gorsuch, Attorney Charles E. Cook filed a suit against the Wright Brothers Company, Wilbur Wright individually, Walter Brookins, and the Aero & Motor Club of Asbury Park, for $15,000 dollars in damages. It was the first lawsuit brought for damages against an aviator. The case was scheduled to be heard at the October Sessions of the Monmouth County Circuit Court.

Two years later, on June 25, 1912, Judge Nelson Y. Dungan ruled in favor of the Wright Brothers. Defense Attorney Edmund Wilson entered a plea that no negligence had been show by the defendants and Judge Dungan agreed. Dungan had no precedent or statute to rule on other than common law. Defendant Aero and Motor Club of Asbury Park was assessed $14.50 court costs. Brookins and the Wright Brothers Company were assessed $55.66 in court costs.

Eighteen-year-old George Burnett, son of Harry B. Burnett of Paterson, New Jersey, suffered a fractured skull, brain concussion and dislocated hip. Onlookers pulled him from under the wreckage, motionless. They put a call into Dr. Edwin Field, the well-known surgeon of Asbury Park. Burnett's injuries were serious. They thought he might not survive and he would be the first fatality. Burnett regained consciousness two days later in the Long Branch Hospital and slowly recovered.

Left, Wright hired Defense Attorney Edmund Wilson. *Courtesy of Red Bank Library. Right, Plaintiff's attorney* Charles E. Cook. *History of Monmouth County, 1922.*

Warren Fromme, son of Isaac Fromme of New York, was also injured in Brookins' crash. The Fromme family was staying at their summer residence in Deal. Others injured included Sam Murgatroyd age twelve of Asbury Park, Gerald O'Grady age thirteen of Paterson, Arthur Healy age thirteen of West Grove, Thomas Woolley age fifteen, and Henry Kruschka of Asbury Park. Two police officers, Richard Brindley and Arnold Reid received cuts and sprains. Brookins said he learned never to try to lift off with the "tail onto the wind."

Governor's Day

Governor's Day was Friday, August 12. It was the busiest day of the event with crowds estimated at 10,000 paying guests. Hundreds of automobiles honked their horns as they watched the planes take off and land. It was perfect flying weather. Margerum scheduled some spectacular events for the day.

Left, US Senator Frank O. Briggs, came out from Trenton for Governor's Day at the Asbury Park Aviation Meet. *New Jersey A History, 1930. Right,* New Jersey Governor John Franklin Fort attended the Asbury Meet. *Library of Congress.*

Left, Captain Claude V. Guerin, veteran of the Spanish-American War of 1898, and Lt. in Co. H., 3rd Regiment, NJ National Guard. *Right,* Asbury Park Real Estate Broker, T. Frank Appleby. *The New Jersey Coast in Three Centuries, 1902.*

New Jersey Governor John Franklin Fort attended the meet on Governor's Day. Captain Claude V. Guerin of Company H, New Jersey National Guard, was a member of the Governor's entourage. Also in special boxes to view the event were Pennsylvania Governor Edwin S. Stuart, Philadelphia Mayor John E. Reyburn, and Asbury Park Mayor Frank T. Appleby. Also in attendance were US Senators Henry Cabot Lodge of Massachusetts, Frank O. Briggs and Senator John Kean of New Jersey; US Congressmen Benjamin F. Howell representing New Jersey Third Congressional District and Henry C. Loudenslager of New Jersey First Congressional District. Also sitting with the Governor were Monmouth County Clerk Joe McDermott, State Prison Supervisor Samuel Kirkbride, Monmouth County Sheriff Clarence E. F. Hettrick and many other local, county and state officials.

Test flights consisted of one or two laps around the field. Margerum planned a contest to see which aviator could land nearest a spot on the field, something Hoxsey accomplished quite easily the day before. They marked out a square with white chalk, one hundred by one hundred feet. Another exhibition flight was designed to show how aeroplanes could be used in war. Hoxsey would fly to about 500 feet, then drop oranges on a model battleship. Another flight was a duration test, and a few other exhibition stunts were planned to round out the program. Afterwards, Governor Fort posed in the biplane seated with Hoxsey. A photograph of the two, Governor Fort in bowtie, Hoxsey in his bold attire, is on page 718 in *Flight*, September 3, 1910.

That afternoon Coffyn went up in an attempt to break the world record for continuous flight. He had been especially trained for this event by the Wright brothers. This was Coffyn's first flight in the eastern states. On his first lap around the field, Coffyn heard a wire snap. He turned off his engine, aborting the attempt, and glided back down.

As the finale event on Governor's Day, Johnny Mack sent up his hot air balloon with two jumpers. The first jump went as planned. The second parachutist, Benjamin Prinz, jumped from the hot air balloon in a stunt with two parachutes. Prinz cut the first parachute away, giving the crowd a thrill, but the second parachute collapsed at about 4,000 feet. Thousands of onlookers looked on in horror as they watched him fall to his tragic death. It was a disastrous event. A young man risked everything and lost his life. Margerum cancelled all hot air balloon stunts for the rest of the meet.

Johnstone's Lofty Flight

Ralph Johnstone desperately wanted to break the world altitude record. *Courtesy of NJ Aviation Hall of Fame.*

The next day Wilbur Wright allowed Johnstone to fly but only up to 5,000 feet. Johnstone had heard that the air was very cold at the higher altitude. Dressed in a woolen winter coat buttoned tight, Johnstone was ready to climb as high as he could, but Wright denied permission. Wright said "If the going up there was just right, I might let him chance it, and if I should, he'd go up to a clean 10,000 feet." Spectators got a dose of "aviator's neck" as they watched Johnstone soar up to an estimated 4,000 feet. His flight lasted thirty-five minutes. On his descent, Johnstone turned off the engine from about 600 feet up and glided into a picture-perfect smooth landing. Despite not breaking the record the crowd mobbed him, men shaking his hand and women hugging him. Johnstone said he had enough fuel to climb higher, but was following instructions from the administration tent.

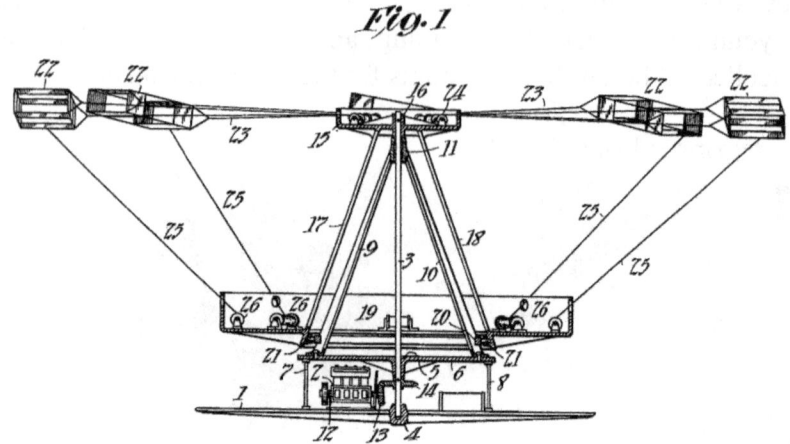

Thomas Edison's "flying machine," Patent No. 970,616, September 20, 1910. *US Patent Office.*

Scientist Day

On Scientist Day, August 16, the Businessmen's Association held a banquet, with Thomas A. Edison guest of honor. They discussed how useful an aeroplane could be during war. Edison had an early interest in human flight. He applied for a patent of his own design for a flying machine in 1908, which the US Patent Office awarded him on September 20, 1910. Edison's design consisted of box kites connected to a frame by piano wires.

Wright's New Five Passenger Plane

The new Wright five-seater aeroplane shipped from Dayton, Ohio, to replace the biplane Brookins demolished, arrived late in the afternoon of August 16. Wright put Coffyn in charge of the new five-passenger plane. Brookins, back on the scene and eager to fly after recovering from his first day crash, wanted to take it up, but winds of twenty-five mph prevented him from flying. Johnstone went up in his biplane but came back after a brief three-minute flight, saying "That was the worst wind I was ever up in."

The Wright brothers' new five-passenger biplane was a radical departure from previous designs. Both Wilbur and Orville disliked the spectacular, stunt flying acrobatics that seemed to thrill the crowd. They envisioned their invention to be used for more practical purposes.

Wright Brothers new 5-passenger biplane, first introduced at Asbury Park Aviation Meet. *Courtesy of NJ Aviation Hall of Fame.*

Tail view of Wright Brothers new plane.

On this new model, Wright removed the two elevating planes from the front and added one elevating plane in the back. This new design eliminated the pitch and roll movement from earlier models. They designed this aeroplane to be a long distance traveling machine, with seats for four passengers and one for the "driver." Wilbur said this model came nearer to his ideal than anything that had yet to be seen in the air.

Brookins with Coffyn as a passenger at Asbury Park. *Library of Congress.*

Coffyn was the first to take up the new Wright five-seater, on the morning of August 18, for a short spin. Later that day, Brookins took it up for a short trial with Coffyn as a passenger. Brookins took it up about 200 feet and did the usual two laps around the field in the air. All went well. Brookins said it felt much faster than the other biplanes. About his new machine, Wilbur said:

> "It takes the quiet concentration kind of work that my brother and I did at Kitty Hawk to advance the progress of the science. While we have never gone in for the spectacular and sensational flying, the kind we did was the most dangerous because we flew in a machine that had never been tried out. This kind of experimenting we expect to continue until we reach a state of perfection with our machines."

Johnstone's Crash

Johnstone had a crash landing on the first exhibition flight scheduled on the afternoon of August 18. His biplane was usually equipped with landing skids but for this flight the skids were replaced with wheels. The Asbury Park Aviation Meet was the first appearance of a Wright biplane equipped with wheels. No one was injured, but his aeroplane crashed through a fence and collided with two parked automobiles. Johnstone's aeroplane engine and wing were destroyed. One car belonged to Henry D. Hackman, a successful cigar manufacturer of Womelsdorf, Pennsylvania. Hackman said he was "honored" to own the first car in a collision with an aeroplane. Johnstone's biplane broke the auto's headlamp and dented the radiator.

Night Flights

On August 19, Hoxsey and Johnstone went up after dark. Promoters had planned to put two aeroplanes up at the same time at night but bad weather forced them to cancel this part of the program. Both aviators were to take a car horn with them, and honk if the other got too close. To show the promoters it could be done, Hoxsey and Johnstone took to the air late on the night of August 19. It was the first two night flights on record. They attached port and starboard lights to their wings. Johnstone went up first in the five-passenger plane. His flight was a short, five-minute circle around the field. He landed at 10:05 P.M. Not to be outdone, Hoxsey took it up shortly after that and did a few figure-eights in the night sky. Hoxsey's flight lasted eleven minutes. Coffyn, who was in charge of the five-passenger plane, did not authorize the flights. He ordered a security guard put at the hangar to prevent the young aviators from taking the plane up without permission in the future.

World Record for Highest Altitude

For three years, 1903-1906, the Wright Brothers held the world record for highest altitude reached in a fixed wing aircraft, at the lofty altitude of ten feet. Wilbur Wright gained back the world record in 1908 when he climbed to 360 feet. A year later aviator Louis Paulhan climbed almost ten times higher, to 3,018 feet. Brookins won a $5,000 prize on June 17, 1910, for setting the record at 4,603 feet at Indianapolis. Three weeks later he broke his own record at 6,275 feet at Atlantic City.

On August 12, J. Armstrong Drexel broke Brookins' world record for highest altitude reached at 6,752 feet. Drexel was from Philadelphia, but set the record in Lanark, Scotland. While recovering from his opening day crash, Brookins learned that Drexel broke his altitude record. He said, "If the Drexel record is made official before the end of the Asbury Park Meet, I will break it at the Interlaken Field," but deteriorating weather conditions prevented his attempt. Both Hoxsey and Johnstone wanted to attempt to take the record back, but Wilbur Wright refused to let them try because only Brookins had been trained to fly at higher altitudes. However, Wright relented.

J. Armstrong Drexel. *Library of Congress.*

On August 20, Hoxsey tried to set the altitude world record mark but failed. His attempt was highly publicized. All seats in the grandstand were sold out. Starting at 4:00 P.M. Hoxsey spent nearly an hour going over his machine while the crowd waited in anticipation. Joseph C. Patterson, Announcer of Special Events, used a megaphone to talk it up, saying Hoxsey was going to shatter the recently set Drexel record. Finally, Hoxsey lifted off the ground to great applause and horn honking. For the first fifteen minutes, Hoxsey climbed in slowly spiraling circles around the field. At about 4,000 feet he leveled off. His goal was to get as close as possible to 10,000 feet.

J. Armstrong Drexel. *Courtesy of NJ Aviation Hall of Fame.*

Before he took off, Henry M. Neely, representative of the Aero Club of Pennsylvania, fixed two barographs to measure altitude onto the wings of Hoxsey's Wright biplane. Dr. Budd H. Obert of the local weather bureau had checked the barographs for accuracy and sealed them. George K. Allen, Jr., County Engineer of Red Bank, was stationed on the beach near the Allenhurst lifesaving station with his clinometer. Allen used treetops to measure and calculate ranges. When Hoxsey landed, they removed the sealed barographs into the administration tent to be read.

Hoxsey reached a disappointing height of 3,458 feet according to the readings. When he landed, Hoxsey said the atmospheric conditions were not favorable. He said the cold air at that altitude chilled him, so he descended. His flight lasted forty-nine minutes. Hoxsey was dressed in light flannels, silk socks and canvas shoes. To some observers, that was not appropriate attire for such altitudes. Some surmised that Hoxsey's gay attire, plus the lack of any prize money, was reason enough to believe it wasn't a true attempt. Promoters for next month's Harvard meet at Boston, offered a $10,000 prize to the aviator that broke the altitude record.

Amateur's Day

Saturday, August 20, was Amateur's Day. The Aero Club offered a fifty dollar award to any amateur aviator who could fly fifty feet powered by their own machine.

Alexander Louis Pepin, an amateur aviator of Asbury Park, planned to fly that day. Pepin was a Canadian immigrant who worked as a shoe salesman. A few weeks earlier, Pepin had purchased a Curtiss engine for the flying machine he had recently built, and planned for its first trial at the Aviation Meet. Pepin arrived the first day of the meet and assembled his craft. Four days into the event, two gears had still not arrived and he could not fly. The design of his craft featured some new and interesting ideas of his own, including three propellers. Several experienced flyers who saw it thought it might be a good design. Reporters called it a handsome machine. Pepin may have had some other trouble because by August 24 he still had not flown. He promised that on August 25 he would be airborne but his wife, Mrs. Emma Pepin, announced he would not fly. Recent accidents at the meet worried her and she was afraid for his life.

Joseph Dunning was the only other aeroplane inventor from Asbury Park who built an aeroplane. Dunning was sixty years old in 1910. He was from New York but lived in Asbury Park and worked as an engineer. The plane he built was very unusual. Dunning claimed his machine could fly 75 miles an hour without rolling over and would not overturn, not even in a strong wind. "It will not turn over" he stated. According to a story in the *Asbury Park Press*, Dunning had perfected his machine after six years in the making. His secret was a six-foot long cylinder which trapped air. Dunning's special device dampened the air in the cylinder which made the air denser, and allowed the propellers to create more speed. A steam engine powered his craft. Dunning invited the experienced aviators to take a look at it, and if they approved, he was willing to arrange a demonstration during the meet.

Dunning's aircraft is pictured in the *Asbury Park Press*, August 14, 1910. As it was true for many unusual designs in the early years of aviation, it is doubtful that this craft ever got off the ground.

World Record for Fastest Turn

Walter Brookins broke his own world record for the fastest 180° U-turn while airborne on August 23, 1910, when he made a full turn in 5½ seconds. It was ¾ of a second faster than his previous record. Designated senior aviator Frank Coffyn was there when Brookins broke the record. Coffyn said:

> "None of the others of us could do it. If we tried, we'd break our necks and we'd know it. Wilbur Wright never got around in the air in less than fifteen seconds, and Orville's fastest time was eleven seconds. How Brookins does it we haven't quite figured out. Until he did it on the new machine, we thought he abandoned the rudder and steered by the front elevating planes. Now we're as much puzzled as before, and we're likely to leave that particular trick to Brookins for his strictly personal use."

Post Asbury Park Meet

As the event drew to a close, the Aero Club removed McCaffrey's canvas wall and tents from the field. Reporters returned to their business offices. The constant parade of Model T automobiles ceased. The landing grounds turned back into a plowed farm field.

Despite one parachutist's death, two aeroplane crash landings, two days cancelled due to bad weather, a failed attempt to break the highest flight altitude record, and a failed attempt to break the endurance record, promoters considered the event a success. The Asbury Park Aviation Meet recorded many firsts in history, including the first time an aeroplane collided with an automobile; the first night flights recorded; the first aeroplane to injure a spectator and ensuing first lawsuit by an injured spectator; the fastest U-turn in midair; the first Wright biplane equipped with wheels, and the first five-passenger plane flown.

Brookins and Johnstone went to Boston for the Harvard meet. At Boston, Brookins set the accuracy record by landing an amazing twelve feet, one inch from the center of a circle. On October 30, Johnstone set the altitude world record by flying 8,471 feet. Two weeks later, Johnstone died in an air crash at Denver, Colorado, on November 17, 1910, holding the world record.

Hoxsey went to the Midwest to perform. On December 26, 1910, Hoxsey broke the altitude record at Los Angeles by flying to 11,474 feet. Five days later, Hoxsey died trying to break his own record while still at Los Angeles on December 31, 1910. Due to the aviator deaths, the extreme danger and Wright's dislike of stunt flying, the Wright Exhibition team disbanded in 1911.

At least one more aviation meet occurred in Asbury Park. During the second week of August 1913, aviator George Gray crashed into a tree near Deal Lake at takeoff. Gray was an Asbury Park resident and expert bowler. He was about thirty feet in the air when his rudder failed. He was badly bruised and taken to the Long Branch Hospital, but returned later that day to watch other flights scheduled during the meet.

That same day, aviator Frederick C. Hild had problems with his plane, when he tangled with some telegraph wires. Hild was an American volunteer for the French military, because "they had the most aeroplanes." He wrote an article for aviation magazine *Sphere*, called "My Experiences as an Air Scout." Hild went on to be the cofounder of Eastern Aeroplane Company of Brooklyn and Aircraft Manufacturing Association of America.

In Memory of Archibald Hoxey (1884-1910).
Courtesy of Empire State Aerosciences Museum.

3. Monmouth Aviation Patents

Thomas M. Walling (1870-1935)

Thomas Malcolm Walling was born on April 22, 1870, in Lockport, a neighborhood in eastern Keyport, Monmouth County, New Jersey, the son of Eugene and Margaret Anne Walling. On Thanksgiving Day, November 30, 1899, Thomas married Elizabeth Cook. By

Thomas Walling, *left,* and his brother Peter, as pictured in *The Globe and Commercial Advertiser,* New York. Photo by George Magee. Courtesy of Dick Winters.

Thomas Walling with his flying machine. *Courtesy of Dick Winters.*

1910, Walling lived on Tinton Falls Road in Tinton Falls, also known as South Shrewsbury, with his wife, three children, his mother and his brother Peter. Thomas worked as a truck farmer. Peter Walling was a civil engineer and an experienced draftsman and construction specialist.

Walling's first full-sized aeroplane. *Courtesy of Dick Winters.*

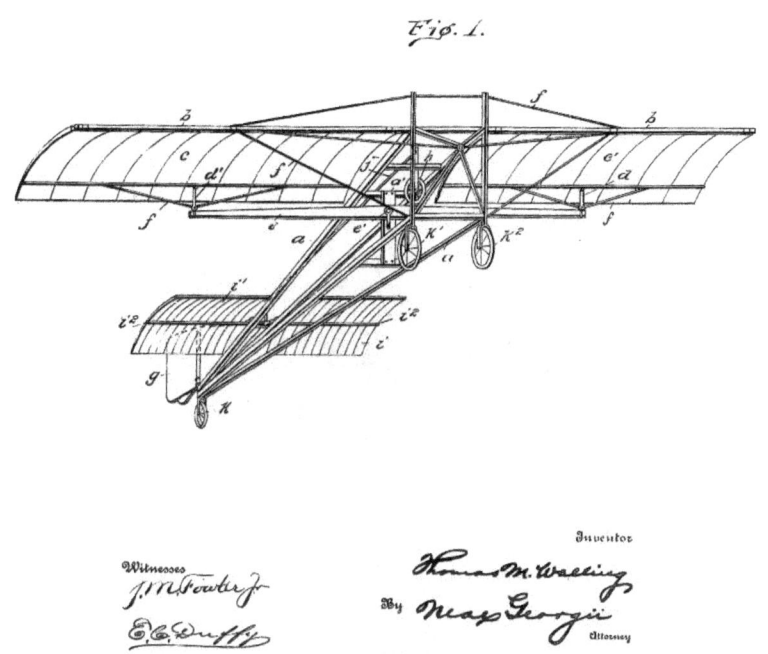

"Aeroplane'" drawing for Thomas M. Walling. Patent No. 1,004,944, October 3, 1911. *US Patent Office.*

Walling's Patent for an Aeroplane

Walling began to design and build aeroplanes in 1908. He worked for two years on his ideas and inventions. Walling was the first Monmouth County resident to patent a flying machine. He filed for a patent on May 20, 1910, for an aeroplane featuring automatic transverse stability. The US Patent Office assigned him application Serial No. 562,409 and issued him Patent No. 1,004,944 on October 3, 1911. Charles Pintard and Albert Van Wickle signed Walling's patent as witnesses to the craft's specifications. Pintard lived in Red

Bank and was a bookkeeper for a plumbing business. Van Wickle was a blacksmith in Eatontown.

Walling incorporated and sold shares to finance his aviation efforts. *Courtesy of Dick Winters.*

Monmouth Aerial

Walling assigned his patent to the Monmouth Aerial and Vehicle Transportation Company, a business he incorporated under the laws of Arizona. As president, Walling issued stock in the company as a way to finance his aviation activities. His brother, Peter, was general manager and his brother-in-law, Joshua Wanhope, was vice-president. Wanhope's wife, Mrs. Sadie Monro Wanhope, was secretary and treasurer. Thomas Walling served as a director along with Mr. and Mrs. Wanhope of New York, and Wellington Wilkins of Tinton Falls.

Wanhope was editor of *Wilshire's Magazine* and also the New York delegate to the Socialist Party of America Convention in 1910. Walling raised money by selling shares of his Monmouth Aerial and Vehicle Transportation Company. Robert Leroy Cook, Elizabeth Walling's brother, purchased 250 shares at $1 per share on August 30, 1910.

Walling's brother-in-law and editor of *Wilshire's Magazine*, Joshua Wanhope helped with some free publicity. *Library of Congress.*

Model Design

Walling said he had solved a stability problem that affected all aircraft. On this patent, Walling claimed his device would automatically stabilize an aeroplane in flight if a gust of wind or machine failure caused the plane to tilt.

Walling made a twelve-foot model and tested it by towing it behind an automobile, and dropping it from a height. Walling claimed that in all tests and under all circumstances, his model righted itself. His model consistently kept an even keel and maintained its balance in the air throughout the flight.

His simple device consisted mainly of a lever attached to what he called an equalizing bar which connected the two main wings of the plane. The two struts were made of spruce, each twenty-one feet long. He positioned the struts one foot apart in the center and clamped them together at the ends with heavy wire.

Once Walling was satisfied with the tests on his model, he began building a full scale aeroplane with the help of his brother Peter.

Walling's twelve-foot experimental model. *Courtesy of Dick Winters.*

Mrs. Elizabeth Walling (1883-1927)

Walling's wife, Elizabeth, was also enthusiastically involved with the project. Mrs. Walling neatly stitched the canvas seams on the wing and rudder, exhibiting her expertise with needlework. When asked about possible danger to the pilot, Mrs. Walling told a reporter:

> "Well I don't know. I'm so enthusiastic over it myself and we have been so busy getting things ready we haven't had time to think about something that might never occur. I have seen a number of monoplanes and biplanes in operation, but I have never yet been afraid that anything might happen to the aviator. We know what the model has done, so we think there is very little danger."

Mrs. Walling was so enthused with flying, she wrote to aviator Ruth Law requesting flying lessons.

Ruth B. Law (1887-1970)

Ruth Law was born in 1887 in Massachusetts and earned her aviator license in 1912. Law replied to Mrs. Walling from the Hotel Clarendon in Sea Breeze, Florida, where she was staying for a few weeks in June 1913. Law wrote back, saying she would teach her but issued her a warning that "there are very few people who are qualified for this profession."

Elizabeth Walling wrote to aviator Ruth Law asking for flying lessons. *Courtesy of Dick Winters.*

In 1916, Law broke the cross-country long distance record by flying 590 miles from Chicago to New York State. After WWI, Law organized the Ruth Law's Flying Circus Act which she booked at state and county fairs to exhibit aerial flying stunts. She briefly held the women's altitude world record at 14,700 feet in 1919. She was the first woman to perform the loop-the-loop stunt. Law's Flying Circus was very popular, but the acts were also very dangerous.

In October 1921, Law was practicing a stunt in Long Branch. A woman passenger in a moving car was to board a low-flying airplane by grabbing a rope ladder dangling from it and climbing aboard. She chose Ocean Avenue in Long Branch to practice because it was well laid out for this purpose. On the afternoon of October 4, Madeline Davis volunteered to get into the car Law was driving. A woman had never done the stunt before, and Davis wanted to be the first.

Davis was an experienced wing-walker. She went to Law's headquarters in Asbury Park and told her "There is nothing on land or in the air that I am afraid of." Law said she would hire her to be part of the Flying Circus if she could do the stunt. They practiced several times along the beach at Long Branch before actually making the first attempt late in the day.

Law drove the car as fast as she could, reaching about 45 miles per hour. Lt. Vernon Treat swooped in going as low and as slow as he could, with the ladder dangling from the plane. The young, aspiring actress reached up and grabbed a rung but as it pulled the daredevil out of the car and she was unable to hold on. Davis fractured her skull in the fall and died trying to perform the stunt.

Law continued to perform the stunt several more times but eventually discontinued doing daredevil acts. One of Law's last contracts was with the Asbury Park Business Merchant's Association in January 1922. Law's star pilot, Verne Treat, performed the loop-the-loop aerial stunt above the Asbury Park Casino to open the merchant's Second Annual Midwinter Exhibit.

Aviator Verne Treat *with hand on propeller,* piloted the plane while Law drove the adapted Special Duesenberg racer. *Courtesy of Missouri State Archives.*

Law retired in March 1922, yielding to her husband's pleas and his fear for her life. She moved to California and lived until 1970.

Monoplane Construction

In October 1910, Thomas Walling rented space in Tilton's Blacksmith Shop in Tinton Falls and spent many hours developing and reviewing the pieces for his automatic transverse stability mechanism. Walling constructed a monoplane using a V-shaped body that had moveable wings attached to the sides. Walling designed the large wings to be easily attached and detached, to accommodate transporting the aeroplane on the ground. The wings also had to be removable because they did not fit through the doors at the blacksmith shop. The machine was twenty-four feet long and thirty-six feet wide. Someone said it looked like an old fashioned cow rack. The tail end of the machine had two small wings. The horizontal tail wing was designed to elevate or lower the plane; the vertical tail wing was used to steer. The propeller was seven feet in diameter with a five-foot pitch. A 35 horsepower motor made by Elbridge Manufacturing Company in Rochester, New York, powered it. In total, Walling's flying machine weighed 600 pounds and had a lifting capacity of 1,000-1,200 pounds. Walling spent many hours and a considerable amount of money for materials on his earlier models and now on this full size plane. He sank his life savings into the project. Without factoring in money for labor, Walling spent $5,000 on materials to build his full-scale aeroplane. Every part was homemade except the motor and the propeller.

Trials

Walling had to sell his house to raise money for his experiments. When he needed more, he sold his horse. Without ever testing at full-scale, Walling believed his automatic stability truss would solve the problems seen on earlier flights. His new plane had many innovative features, but he had not patented all of them. Walling was very confident that all his innovations would work. Support came from Lewis S. Thompson, who offered Walling his forty-acre field in Brookdale to conduct the test flights.

By September 28, 1910, Walling announced he was ready for a trial run but that trial had to postponed due to receiving a motor that would not operate the propeller. He returned the motor and had a different model shipped. A week later, his new 40 horsepower motor arrived. Walling connected it to the propeller and, in front of a crowd of about seventy-five people, started the engine.

Retired Sea Captain David A. Walling opened a general store in "Shoreville," now South Belmar. *History of Monmouth County, New Jersey, 1922.*

Walling tested his new engine at Captain David A. Walling's cider mill and distillery in Tinton Falls. Before starting, he anchored the aeroplane to the ground with ropes so that it would not get airborne. This time the engine worked perfectly and so did the propeller. Several witnesses said the plane seemed to be ready to fly and appeared to be well built and very sturdy, but Thomas Walling wanted to strengthen the machine with more wire and more braces before he was ready to leave terra firma.

In early November 1910, Walling announced he was ready to give his machine its first trial flight. Walling was satisfied with the machine's strength. The motor and propeller were working in unison. The automatic stabilizing bar was working properly. He said he had low expectations for the first flight, and if it hopped along the ground for ten to twenty feet he would be pleased. That would give him the opportunity to learn more about how to handle the machine at higher altitudes. Andrew R. Colman, photographer for the *Red Bank Register*, took a few pictures which were published in the edition of November 2, 1910.

On Friday afternoon, November 19, Walling hauled his craft to Thompson's field. Some said it flew a short distance. Walling made a few more adjustments and was ready to try it again the next day, but problems with the engine failed to get the craft airborne. He tried again on Monday with the same unsuccessful results. Walling considered having a company expert available to help adjust the motor settings for the next trial but eventually concluded that his airship was too heavy to fly.

Thomas and his brother spent that winter building a second model, which weighed half as much as his first airship. He and Peter manufactured several hundred steel turnbuckles by hand in the blacksmith shop. They continued to sell stock in the company to raise funds. A reporter wrote that a "considerable amount of stock was sold to people in the area who had faith in Walling's inventive genius."

Walling exhibited his new aeroplane on September 1, 1911, in a forty by eighty-foot tent near the bleachers at the Monmouth County Agricultural Fair in Red Bank. Whether Walling ever got airborne in the second flying machine is unknown.

In the early 1920s, Thomas Walling sold his farm in Hazlet, bought a new Ford touring car, and moved to Ocala, Florida, where he bought another farm. He died on December 22, 1935, of acute heart failure on his farm in Santos near Ocala, and is buried in the Walling family plot in Atlantic Cemetery in Colts Neck, New Jersey.

Although there was a Walling Airport in Keyport, Thomas and Peter Walling were not part of it. It was owned by Daniel Walling and his brother James. The Walling Airport was formerly called the Keyport Airport as early as June 1929 when Lt. Commander George R. Pond and L. V. Rawlings flew from there, in a Fairfield monoplane to the steamer *Leviathan*, 60 miles at sea.

Clarence L. Moore (1888-1963)

Clarence Levi Moore was born on December 21, 1888, in Wilson, Niagara County, New York, the son of Thomas Moore, a Canadian immigrant and newspaper dealer. He worked in the machine shop for the Merritt Manufacturing Company in Lockport, New York, before moving west to Seattle, Washington, in 1916. There he registered for the WWI draft in 1917. In 1919, Clarence married Christine Evenson, a Norwegian immigrant, and by 1920, Moore was working as an aeroplane mechanic in Plant No. 2 at the Boeing

Aircraft Company. Clarence, or "CL," as his friends called him, was an inventor. Besides his aeroplane patent in 1911 he also invented a pipe wrench, in 1920 under US Patent No. 1,404,178.

An avid fisherman according to family lore, CL designed several fishing lures, which he may or may not have patented. Moore lived longer than his four brothers who stayed on the east coast. He retired from Boeing in 1956 and died April 7, 1963, in Seattle.

Clarence L. Moore of Allenhurst patented an aeroplane. *Courtesy of Kris Davis.*

Moore, *standing second from left,* with his arm around (his brothers?) at the Merritt Machine Shop, Longport, New York. *Courtesy of Kris Davis.*

Moore Patent for an Aeroplane

On September 9, 1911, Moore filed an application for a patent on an aeroplane of his own design. The US Patent office assigned his application Serial No. 648,472. At the time he filed, Moore lived in Allenhurst, Monmouth County, New Jersey. Moore spent only a few years in Allenhurst. In 1910, he was living with his uncle in New York, and by 1916 he had relocated to Seattle. Allenhurst may have been a summer residence for the Moore family, or Clarence could have been attracted to the area to take advantage of good fishing or flying opportunities.

To process his application Moore hired Attorney William T. Jones, of Washington, DC, who specialized in counseling clients for patent and copyrights.

In 1908, Jones published a forty-eight page "Patent Manual for Inventors and Manufacturers." Jones charged twenty-five dollars for his service as a patent attorney. The US Patent Office required an initial filing fee of fifteen dollars for a patent application and a final fee upon awarding the patent of twenty dollars. Drawings cost five dollars per page and Moore submitted three drawings. His patent application cost him seventy-five dollars. Moore probably chose to hire an experienced patent attorney like Jones in Washington, DC, instead of a local lawyer, because of his serious belief in his design.

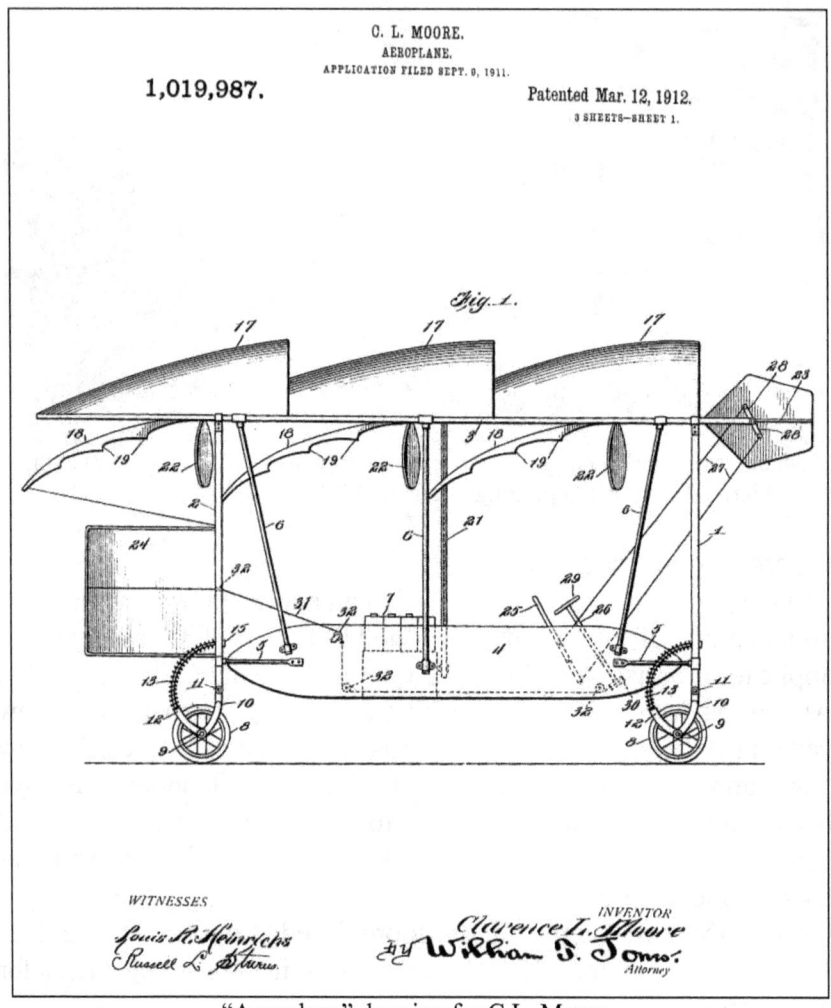

"Aeroplane" drawing for C.L. Moore.
Patent No. 1,019,987, March 12, 1912. *US Patent Office.*

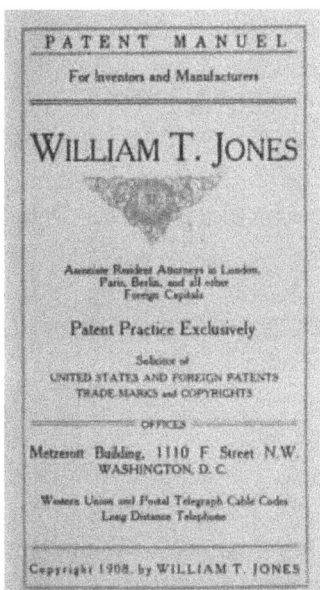

Front cover of Patent Attorney William T. Jones' "Patent Manual for Inventors and Manufacturers," Washington, DC, 1908. *Library of Congress.*

Louis R. Heinrichs and Russell L. Stevens signed as witnesses to the drawings on Moore's application. Heinrichs and Stevens both lived near and worked in Washington, DC, and were most likely acquaintances of Jones. Heinrichs was a draftsman who probably sketched the patent drawings of Moore's aeroplane design. Stevens was a law office clerk who went on to become a patent attorney. The United States Patent Office granted Moore his Patent No. 1,019,987 on March 12, 1912.

Moore admitted his design was unusual. He referred to it as "a novel construction and arrangement of sustaining elements." Moore's design featured three aerofoils which he claimed would promote lifting power of the machine. In the event of an engine failure, the aerofoils would serve as a parachute to ensure the slow and safe descent of the machine to the ground. His alignment of the hood-shaped aerofoils was designed to create stability and decrease or eliminate sideways drift or skidding. A shaft connected a series of three rotating propellers designed to force air into the aerofoils.

Clarence Clayton and David Thurston signed as witnesses to the specifications and claims on Moore's patent. Both Clayton and

Thurston lived in Asbury Park, adjacent to Allenhurst. Clayton was born about 1866, in New York. He lived on Bond Street and worked as a bookkeeper for a printing office. Still single in 1911, he lived with his cousin Harry W. Howland and his family. Thurston was also from New York and about the same age as Clayton. A wealthy real estate broker, Thurston lived just a few blocks away on Grand Avenue with his wife Emma.

Whether Moore ever actually built a machine according to his patented design is unknown. Family descendants were aware of his tool patent but not of him building an aeroplane.

William C. Hurst (1867-1928)

William Charles Hurst was born in Crosby, Lancashire, England, about 1867. Hurst was granted two aviation patents. Hurst came to New York with his parents before he was ten years old. He married Johanna C. Oettinger on April 15, 1889, in New York City. In 1900, Hurst worked as an electrician. In 1910, he lived in the Bronx and worked as a chauffeur. He eventually moved to Bradley Beach. He died October 25, 1928, in Manhattan and is buried in Green-Wood Cemetery in Brooklyn, New York.

Hurst Patent on Airship with Moveable Grouped Ailerons

William Hurst first applied for a patent on an "Airship" on December 1, 1909. The US Patent Office issued him Patent No. 994,104 on May 30, 1911. Witnesses to Hurst's specifications in 1909 were Henry J. Lucke, and George N. Kerr. Harry J. Lucke was a patent attorney of East Orange, New Jersey. Lucke also witnessed the drawings, along with A. K. Schneider. Hurst hired the attorney firm of Edwards, Sager & Wooster. Attorneys Clifton V. Edwards, Lawrence Sager, and Julian Wooster, had experience representing clients in patent infringement cases.

Hurst's design incorporated eight moveable, triangular-shaped wings, made of aluminum or aluminum alloys. His moveable wings allowed the rudder to be smaller in length and lighter in weight than previously designed aircraft. The groups of wings were mounted parallel to each other and arranged horizontally one above the other. The wings were mounted on shafts that could be turned. A series of levers connected the shafts to allow the pilot to move a set of wings simultaneously for uniform positioning if desired. His drawing showed two four-cylinder engines, each powering a separate propeller. Hurst included two parachutes as part of his design.

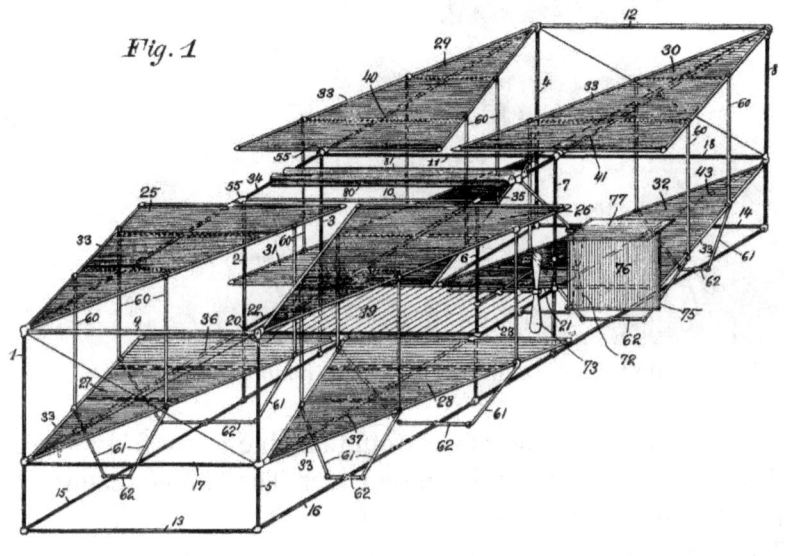

"Airship" drawing for William C Hurst. Patent No. 994,104, May 30, 1911. *US Patent Office.*

Moveable Wings on Hydro-Aeroplane

Hurst hired Aeronautical Engineers Charles & Adolph Wittemann of Staten Island to build the plane according to his design. Hurst designed a new type of aircraft that was a combination biplane hydroplane. He wanted it to be able to carry a pilot, two passengers,

Charles and Adolf Wittemann. *Courtesy of NJ Aviation Hall of Fame.*

and a machine gun. The plane would be able to fly forty to eighty miles per hour and touch down on water. The unique feature of his biplane hydroplane was its moveable wings.

The pilot had controls to manipulate the wings and each wing could be moved independently. One wing could be lowered while the other was raised, in order to maintain a level balance, much like a bird used its wings to maintain balance in flight. Hurst's plane had a wingspan of forty feet.

The Wittemann brothers made a plane that was about twenty-five feet long. A 100 horsepower eight-cylinder Curtiss engine powered an eight-foot propeller. The frame was entirely of steel construction and it sat on two floatable pontoons for landing. Hurst called it the *George Washington*.

The *George Washington*

On November 5, 1914, Harold W. Blakeley tested Hurst's new plane off of Oakwood Heights on Staten Island, New York. On its maiden test, the plane traveled 1,000 feet on the water, then rose to a height of about forty feet, then dove down and crashed into the Lower Bay. Blakeley was thrown from the plane but unhurt. It was the first time Blakeley had flown the plane. He admitted control was a bit different and would take some getting used to. Blakeley said the craft had excellent lifting capacity allowing it to reach its altitude easily. The tip of the right wing was damaged, but not enough to postpone a second test flight that afternoon.

Blakeley was an expert aviator and flight instructor. He piloted many test flights but died a tragic death. On January 16, 1918, Blakeley was testing a new type of Liberty motor in a biplane at Central Park, Long Island, when the plane crashed from about 200 feet. He died on the way to the hospital.

US Navy Wants Seaplanes

William C. Hurst was among fifteen bidders in February 1915 when the US Navy requested bids for six military hydro-aeroplanes. Hurst bid $11,000 per plane but was outbid by several other companies. The Navy wanted biplanes of the seagoing type that could carry two persons, guns, ammunition, and armor, for a total load averaging 600 pounds. They were required to travel fifty to eighty miles per hour, and weigh between 2,500-3,000 pounds. Fifteen companies sent in bids. The Navy refused to divulge any information from the bids, officially stating that "every effort will be made to keep certain improvements devised by some of the bidders from becoming public."

Hurst Aircraft Corporation in Bradley Beach

On February 14, 1917, William C. Hurst, of 506 Third Avenue, Bradley Beach, filed a Certificate of Incorporation with the Monmouth County Clerk's Office in Freehold, New Jersey, for "The Hurst Aircraft Corporation." Hurst named Bernard V. Poland, Esquire, as agent, whose principal office for the firm was located in the Appleby Building in Asbury Park. He formed the corporation "to build, construct, equip and reconstruct aircraft, aeroplanes, or parts thereof of the Hurst Patent No. 994,104 and all other improvements, or patents relative thereto to sell, lease or otherwise dispose of same, to build all necessary buildings and purchase all necessary land, and to do any and all things necessary in connection with the business of the corporation." He issued 1,250 shares at par value of $100 each and authorized $125,000 capital stock.

Hurst issued shares to three partners. He named Ivar Andreas Bjornstad as secretary, and issued him four shares. Bjornstad, of 506 LaReine Avenue in Bradley Beach, was a civil engineer born in 1873, in Norway.

Hurst also issued four shares to his partner Thomas A. Buckingham and named him treasurer. Buckingham was born about 1875 in Canada and ran a stationery store at 716 Main Street in Bradley Beach.

His third partner was Richard Edgar Adamson DeBow of 107 Third Avenue in Bradley Beach, who received just two shares. DeBow was born on August 18, 1887, in Philadelphia. In 1917, he was a self-employed general contracting engineer.

As president of his own company, Hurst planned to buy property on Shark River and build a factory at Bradley Beach. Despite his plans, The Hurst Aircraft Corporation never purchased any property in Monmouth County, New Jersey.

In July 1920, *Aerial Age Weekly* reported that E. H. Snyder, vice-president of the Hurst Aircraft Corporation was accepting bids for building a factory on the fields in Evansville, Indiana. The company had been perfecting a three-passenger plane that would be economical and durable. The new Hurst Aeroplane Company President, M. V. Yeager, and Treasurer Will O. Ferguson, purchased a brick building to temporarily use as a laboratory and assembly plant in Indiana.

Hurst Patent for Aeroplane Steering Wheel

On May 23, 1921, Hurst filed a second patent, this one for a "steering wheel for aeroplane." The US Patent office granted him Patent No. 1,424,143 on July 25, 1922. Patent attorney Henry J. Lucke was the only witness that signed his application.

"Steering Wheel for Aeroplanes" for W. C. Hurst. Patent No. 1,424,143, July 25, 1922. *US Patent Office.*

Stephen F. Stevens (1885-1965)

Stephen Forge Stevens was born on November 14, 1885, in Rumson, New Jersey, the son of a school janitor, George W. Stevens. Stephen Stevens worked as a local building contractor. On April 18, 1928, Stevens filed an application for a patent for "parachute equipment for aeroplanes." He hired attorney Clarence A. O'Brien to represent him.

"Parachute equipment for aeroplanes," drawing for Steven F. Stephens of Rumson. Patent No. 1,702,422, February 19, 1929. *US Patent Office.*

Stevens' invention was a safety device, designed to permit the slow descent of a disabled aeroplane as it floated down to earth upon engine failure. Stevens designed a pocket added to a section of the wing that stored a tank of helium gas and a large folded parachute. When needed, the aviator could pull a lever and activate the device.

The parachute would float 1,600 pounds, the average weight of an aeroplane, but larger parachutes for heavier machines could be used. The parachute material would be made of buoyant material, also serving as a life raft and preventing the aeroplane from sinking if it landed over water. Stevens built a five foot by one foot working model and stored it at his house.

Stevens also invented an escape tank for men trapped in a sunken submarine. He died in Red Bank on March 21, 1965.

Ralph P. Fox (1886-1978)

Ralph P. Fox was born in 1886, in Lebanon, Pennsylvania. He enlisted in the US Army on June 13, 1911, at Fort Slocum, New York, and was assigned to the Coast Artillery Corps. In 1913, while stationed at Fort Hancock, Sandy Hook, New Jersey, Fox filed for a patent on a "safety-support for flying machines" which was awarded that same year.

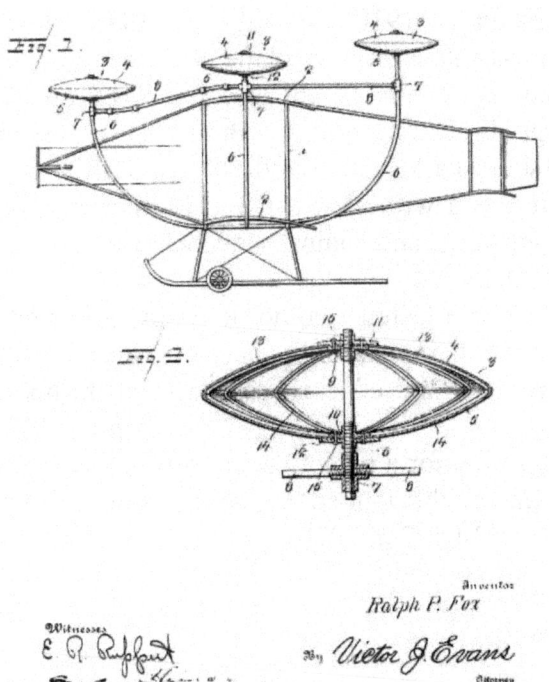

"Safety support for flying machines" drawing for Ralph P. Fox of Fort Hancock. Patent No. 1,073,977, September 23, 1913. *US Patent Office.*

Three witnesses signed Fox's application: telegraph operator George W. Hardy of Atlantic Highlands; US Government Guard Howard A. Johnson of Atlantic Highlands; and E. R. Ruppert. Fox hired patent attorney Victor Justice Evans of Washington, DC, to process his application.

Fox claimed his safety-support invention would provide balance and support for an aeroplane. Six concave balancing structures, triangular in shape and made of canvas, were attached to the rear and sides of the aeroplane. Fox applied for his patent on March 7, 1913, and received patent No. 1,073,977 on September 23, 1913. It took the US Patent Office just six months to issue, which in those days was quick. His application may have been expedited due to Fox being in the service. Fox received an honorable discharge June 12, 1914.

J. Thomas Havens (1855-1929)

John Thomas Havens of 605 Bangs Avenue, Asbury Park, worked as an undertaker in 1910. On July 30, 1915, Havens filed an application for a patent on "aeroplane guy wire tighteners." The US Patent Office issued him Patent No. 1,165,168 on December 21, 1915. George L. Blume and C. Bradway witnessed his patent drawings. Asbury Park Justice of the Peace Martin L. Ferris signed his application as a witness. Scott R. Haycock also witnessed his patented description and claim. Haycock was a draftsman from Washington, DC.

Havens' guy wire fastener could be easily tightened or released. He claimed his device was simple and inexpensive to construct and that it was reliable and efficient to use in all forms of aeroplane construction.

Havens also filed for a patent on improvements in aeroplane guy wire tighteners in the United Kingdom, and was issued British Patent No. 191,514,497 in 1916.

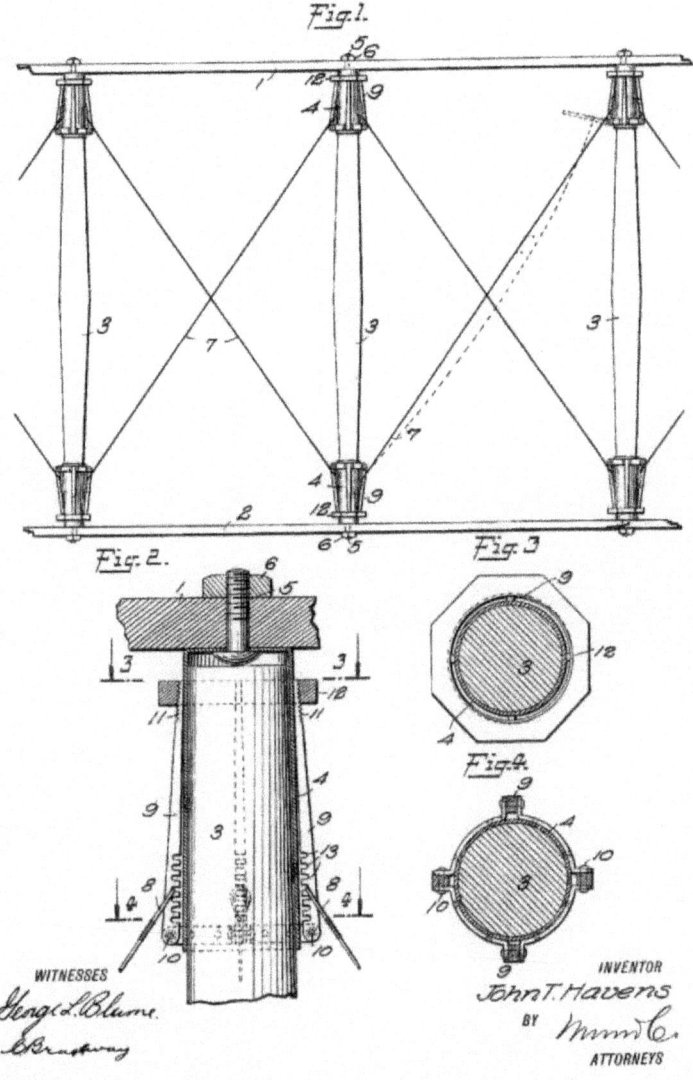

"Aeroplane guy wire tightener," drawing for J. Thomas Havens of Asbury Park. Patent No. 1,165,168, December 21, 1915. *US Patent Office.*

4. Gala at Rest Hill

"The flying machine should be unselfishly and rapidly developed to its ultimate potential for economic advancement in America."
Robert J. Collier

Passport Photo Robert J. Collier, January 4, 1916. *National Archives.*

Robert J. Collier (1876-1918)

Robert Joseph Collier was born on June 17, 1876, in New York City, son of Peter F. Collier. He graduated from Georgetown University in 1894, and took over as publisher of his father's *Collier's Weekly* magazine. An avid sportsman, Collier enjoyed fox hunts, big game hunting in Wyoming, polo and aviation. Robert Collier didn't build an aeroplane or patent a design. The wealthy publisher was fortunate enough to pay designers and builders to make aeroplanes according to his preferences. He hired experienced pilots and mechanics as personal assistants. Collier bought several planes during his lifetime. In 1911, Collier was elected president of the Aero Club of America. In 1912, he wrote an article for their bulletin called "The Alluring Sport of Flying", and began sponsoring the Aero Club Trophy. Awarded annually for the greatest achievement in aviation, it has since been renamed the Robert J. Collier Trophy in his honor. The massive original Collier trophy is on permanent display at the Smithsonian Institution National Air and Space Museum in Washington, DC.

Robert Collier died November 9, 1918, of a heart attack and is buried in the Collier Family Burial Ground at his Rest Hill Estate in Wickatunk, New Jersey. He was inducted into the New Jersey Aviation Hall of Fame at Teterboro in 1984.

The *Laredo*

Robert Collier became fascinated with flying shortly after the Wright brothers' successful flight. Convinced that flying was the future, Collier became a financial backer of Wilbur and Orville Wright when they incorporated as the Wright Company on November 22, 1909. By 1910, Collier traveled to Augusta, Georgia, to become one of Frank Coffyn's first students at the Wright Flying School.

In January 1911, Collier purchased a Wright Model B aeroplane from his friend Orville Wright. It was the first Wright machine sold to an individual for pleasure and the first to be produced in quantity by the Wright brothers. The Model B was a two-seater biplane. Collier reportedly paid $5,900 for his Wright biplane, and rented a hangar at Belmont Park. Collier made a few flights in it at Wickatunk, then loaned it to the War Department. The US Army took delivery at Laredo, Texas, in March 1911. Collier's *Laredo* had a maximum speed of forty-five mph and a range of 110 miles. When the encampment at Laredo closed, they shipped Collier's *Laredo* back to his Rest Hill Estate in Marlboro.

Oliver G. Simmons, left, in Collier's *Laredo* Wright biplane; photo by James H. Hare of *Collier's Weekly*. *Courtesy of NJ Aviation Hall of Fame.*

Youthful Collier, *left,* seated with elder Henry Lloyd Herbert, great polo enthusiast and close friend of Collier's father, photo James H. Hare of *Collier's Weekly*. *Courtesy of Sister Debby Drago, Jean Navagh, and Dorothy Herbert Yeargin.*

Rest Hill Aviation Gala

Starting October 11, 1911, Collier hosted his own five-day aviation meet at his palatial Rest Hill estate in Wickatunk. Hundreds of local farmers came to enjoy the event. The press called it "the greatest event in the history of Wickatunk." Collier invited all the local farmers and several well-known aviators. He offered dozens of flights daily, a polo match and a fox hunt. Workers set up four large food tents and Collier supplied a small army of waiters to cater to his guests. Reporters estimated over 3,000 guests attended. One of Collier's pilots was Oliver G. Simmons.

Oliver G. Simmons (1881-1948)

Oliver George Simmons. *Courtesy of NJ Aviation Hall of Fame.*

Oliver George Simmons was born in Philadelphia, about 1881. Simmons was a veteran of the Spanish-American War of 1898. In 1902, he re-enlisted in the US Army Signal Corps, and served as a sergeant in charge of the engineering department. By 1911, Collier had hired Simmons as an aeroplane mechanic, then retained him as his private aviator.

On the first day of the Gala, Simmons made ten successful flights for Collier's guests. Simmons took photographer William Manna up as a passenger to take photographs of the countryside. He also flew with passenger Henry Conover, age thirteen, son of Charles E. Conover, who might have been the youngest passenger to ever fly.

Simmons gave passenger Monmouth County Clerk Joe McDermott a flight that lasted twenty-five minutes.

On the last day of the affair, Collier and Simmons were in his *Laredo* about 1,000 feet in the air near Lakewood when the engine died, and they had to glide down to land. Collier had arranged for his guests at Rest Hill to travel in automobiles and planned to meet them at DeLisle's at Allaire for a luncheon.

With his stalled engine, Simmons aimed for a field on James W. Wagner's farm. As they neared the ground, one wing struck some corn stalks, causing the aeroplane to flip over. Collier leapt out of the craft just as its nose dove into the ground. He was badly shaken but not injured. Simmons suffered severe cuts. They had flown about 35 miles but now were forced to make the last leg of the trip in a horse and wagon, driven by Wagner. Despite the narrow escape, Collier offered Simmons a job as his personal pilot at Wickatunk which he accepted. Simmons had designed and patented an aeroplane engine. He designed many machine and tool parts and was credited with obtaining more than twenty patents.

Simmons, *left,* with South Amboy Mayor Frank Garretson. *Courtesy of NJ Aviation Hall of Fame.*

On July 4, 1912, while still employed as Collier's personal aviator, Simmons made a historic flight from Perth Amboy to South Amboy carrying the US Mail. The South Amboy Businessmen's Association sponsored the event as part of an Independence Day celebration. The US Postal Service agreed to authorize the flight as long as the pilot and passenger were sworn mail carriers and the flight would incur absolutely no cost to the Postal Service. Collier agreed to lend his Wright-Burgess hydro-aeroplane free of charge.

South Amboy Mayor Frank Garretson accompanied Simmons as a passenger holding the bag of mail. It was a short flight, but it was the first authorized delivery of air mail in New Jersey. Simmons flew across Raritan Bay in Collier's hydro-aeroplane. The event was highly publicized. It attracted crowds at takeoff in Perth Amboy and more crowds at the landing field in South Amboy. Simmons made several more flights that day to the delight of the crowd but without carrying mail. The South Amboy Businessmen's Association presented him with a silver loving cup.

Simmons Cup, in honor of the first US Air Mail delivery in New Jersey, July 4, 1912.
Courtesy of NJ Aviation Hall of Fame.

Simmons Supports Woodrow Wilson

At the Keansburg Carnival in August 1912 Simmons made exhibition flights in the hydroplane to thrill the crowd. Simmons picked up passenger James Hubbard at Seidler's Beach and rose to about 200 feet. Hubbard had political flyers supporting Governor Woodrow Wilson in his bid for presidency which he and Simmons threw from the plane. While they distributed the flyers, the plane suddenly nosedived.

> **WOODROW WILSON**
>
> September 5, 1912.
>
> Dear Mr. Simmons:
>
> I was indeed sorry to learn that you had met with a serious accident in Sandy Hook Bay on Saturday last, and desire to express to you my sincerest sympathies, and the hope that you will soon be well again.
>
> I have learned of your kindly interest in me and my candidacy, and I wish to express to you my appreciation of your support and good feeling.
>
> Cordially yours,
>
> Woodrow Wilson
>
> Mr. O. C. Simmons,
> Wickatunk, N. J.

Letter from Governor Woodrow Wilson to Simmons, September 5, 1912. *Courtesy of NJ Aviation Hall of Fame.*

Simmons was able to glide down but the plane crashed in Raritan Bay, about 150 feet off the Keansburg Boardwalk. Simmons wore a life preserver and escaped major injury but Hubbard did not.

Volunteers from the Keansburg Life Saving Crew rescued Hubbard, who was pulled from under the sinking plane. Hubbard suffered a head injury, a cut at the ear and a bruise on his back. They brought him to Dr. Edwin Field's office in Red Bank where he was treated. The crash landing smashed the plane badly rendering it unusable. Simmons suspected the sudden fall was due to a broken propeller shaft.

With the Monmouth County Fair just four days away, Simmons quickly contacted Collier who was with Frank Coffyn, flying Wright hydroplanes at Raquette Lake, New York. Collier had a plane shipped in time for Simmons to fly at the Monmouth County Fair, where presidential candidate Woodrow Wilson was scheduled to attend. Simmons, still employed by Collier, gave more exhibition flights during the first three days of the Fair. Wilson campaigned at the Fair, then won the election by a landslide. Simmons was inducted into the New Jersey Aviation Hall of Fame in 1987.

Aerial view of the airfield and polo ground at Rest Hill, Wickatunk. Hangars, *left*, Collier's estate residence *at right*. *Courtesy of NJ Aviation Hall of Fame.*

Aviator Thomas Sopwith. *Library of Congress.*

Thomas Sopwith (1888-1989)

Collier invited Thomas Sopwith to fly at his Gala. Sopwith was born in Kensington, London, on January 18, 1888. The English ice skating champion taught himself how to fly at a young age. He had recently won £4,000 for being the first aviator to fly from England across the English Channel to the continent in an English-made plane. Sopwith flew from Eastchurch, England, and landed in Beaumont, Belgium, on December 18, 1910, just two months after his first solo flight, in a Howard Wright Avis monoplane. Sopwith used his winnings to finance a trip to the United States and to participate in exhibitions. He and his sister Margaret, whom he had trained as an aviator, came to Philadelphia aboard the steamer S.S. *Amerika*. He shipped his 70 horsepower Bleriot monoplane via the

French liner S.S. *La Provence*. Sopwith participated in the Philadelphia exhibit, then flew at Mineola in New York.

Sopwith. *Library of Congress.*

From there he shipped his plane to Wickatunk, Monmouth County, to be a part of Collier's Gala. At Rest Hill, aviators Sopwith, Simmons, and Al Welsh, made many flights, taking up passengers each time. Sopwith's Bleriot monoplane had engine trouble, which prevented him from flying in his own monoplane. He used Collier's Wright biplane to make the flights.

Panama

In January 1912, Collier sailed aboard the S.S. *Turrialba* to Colon, Panama, in pursuit of a $3,000 reward for the first aviator to fly in Panama. Collier brought with him his aviator Al Welsh, mechanic Oliver Simmons, *Collier's Magazine* photographer James H. Hare, and his Wright Model B "*Laredo.*" Hare was a veteran war photographer and Collier's personal photographer. Hare traveled to Cuba to photograph the *USS Maine* and life in Cuba after the Spanish-American War of 1898.

In Panama, Collier visited the Atlantic Coast and the Panama Canal but was unable to find a suitable takeoff and landing site. He quickly discovered that the moist tropical climate caused problems with flying machines. High humidity loosened the glue used to hold the propellers together and the rubberized wing coatings were not effective in creating lift. When Simmons told Collier that he could only expect a maximum fifteen-minute flight, Collier abandoned the attempt and the foursome returned home.

Alfred L. Welsh (1890-1912)

In 1890, Alfred L. Welsh (aka Al or Arthur) was a Russian-Jewish immigrant to America at nine years old. Wright employed him as part of the Wright Exhibition Team. He soon became an instructor for the Wright Brothers Flying School. In 1911, Welsh won the endurance prize in a Wright machine at Chicago by being the first aviator to travel for two hours with a passenger. He was also successful at setting other aerial records for speed.

Aviator Al Welsh traveled to Panama with Collier. *Library of Congress.*

A few months after returning from the Panama trip with Collier, Welsh died in an aeroplane crash on June 11, 1912, at College Park,

Maryland. With passenger US Army Lt. Leighton W. Hazelhurst, Jr., Welsh was trying to complete a War Department requirement of being able to climb to 2,000 feet in ten minutes with a load of 450 pounds. At about seventy-five feet, the wings detached and the plane dove to the ground. Both pilot and passenger were killed on impact. Welsh was the first American Jewish aviator.

Collier's Aero-skimmer

In the spring of 1913, Collier purchased two more planes. He purchased a Sloane 220 horsepower Aero-skimmer made by the Sloane Aeroplane Company of New York, and a Burgess hydro-biplane. Collier hired aviator Frank Coffyn to design the Aero-skimmer according to his specifications. Collier wanted it to be capable of flying 60 miles per hour with a seating capacity for six people. He wanted it to be able to carry enough fuel for four hours flying time with one or two passengers aboard. Based on Collier and Coffyn's design, John E. Sloane built a gliding-boat that looked like a huge bobsled. The engine Collier selected and had installed made it the most powerful machine ever constructed of its type, enabling it to glide swiftly over the surface of the water. The hulls floated in just sixteen inches of water. Such shallow draft allowed the craft to land and takeoff on remote streams. It would be a boon for sporting purposes.

John E. Sloane built a hydroaeroplane for Collier.
Library of Congress.

Sloane, Coffyn and Collier's final design was eighteen feet, three inches in length and thirteen feet wide. The hull consisted of five very wide hydroplaning surfaces, attached to a girder framework, one behind the other, six inches apart. This allowed minimum drag and maximum planing surface. Two centered main beams carried six seats for the operator and the passengers directly in the center of the machine.

Sloane installed an Anzani motor made by the Alessandro Anzani Company of France. The 220 horsepower, twenty-cylinder, air-cooled, Anzani motor drove the eight foot, four-bladed propeller. The motor alone weighed over 700 pounds. The engine was too difficult to crank by hand so Sloane installed a Hartford reducing gear with a push button starter. Several experienced flyers thought the push button starter was quite impressive. Push a button and start your aeroplane - the epitome high tech aviation luxury circa 1913! The pilot operated the rudder by an automobile steering wheel located at the extreme front of the hydroplane. Sloane produced several more of his Aero-skimmers.

Aviator Frank Coffyn. *Courtesy of NJ Aviation Hall of Fame.*

Frank Coffyn tested the new plane for the first time on July 19, 1913. After a few trial runs to get the feel of the machine, Coffyn took it up. The machine reached a speed of seventy-five mph, much to the delight of Collier, Sloane and Coffyn.

Collier's Burgess hydro-aeroplane, docked in Raritan Bay. *Courtesy of NJ Aviation Hall of Fame.*

Collier's hydro-biplane

Collier's other purchase in 1913 was a Burgess hydro-biplane made by the Burgess Company and Curtiss in Marblehead, Massachusetts. The Burgess was much bigger than the Aeroskimmer. The warping upper wing spanned forty-one feet and 4½ inches. It was made in four sections, with five foot wing tip extensions. The lower wing was rigid and built of tubular steel. Minus the five foot extensions, the lower wing spread thirty-three feet, 4½ inches. With wing depths of five feet, six inches, it had 373 square feet of supporting surface, braced by wood ribs spaced twelve inches apart. The overall length of the plane was thirty fee,t six inches. The twenty-eight foot long hull was of solid mahogany and contained tanks for gas and oil. The motor weighed 968 pounds and powered a four-bladed Burgess-type propeller, eight feet in diameter, with a pitch of seven feet, nine inches. Collier paid $10,000 for the Burgess and bought the motor separately.

Test flights off Cliffwood Beach

On December 23, 1913, Coffyn made four flights off Seidler's Beach. Located within 50 miles of New York City, Frank Seidler catered to airmen. Coffyn took a passenger on each flight. He flew at varying altitudes. Some observers estimated he reached eighty mph. After the wintery test flights, Collier ordered both hydroplanes dismantled and stored until spring. Well known aviators that stayed at Seidler's Beach Hotel included Collier, Brookins, Coffyn and several others.

SEIDLER'S BEACH
NEW JERSEY

is the first bonafide flying boat station in the world.

It is the most ideal landing place for flying boats within 50 miles of New York City.

We Cater to the Airmen

No owner or pilot of either a flying boat or hydro-aeroplane should come to New York without visiting Seidler's Beach.

Seidler's Beach Hotel is especially adapted for flying men. Such well known flyers as Robert J. Collier, Walter Brookins, Frank Coffyn, Grover C. Loening, Alfred W. Lawson, Walter E. Johnson, Earl Beers and Barton, have made this beach their headquarters.

For Further Particulars Communicate with

FRANK SEIDLER
CLIFFWOOD - NEW JERSEY

Seidler advertised his hotel on Raritan Bay as "the first bonafide flying boat station in the world." *Aircraft, November 1913.*

5. Aeromarine Plane and Motor Company

Joe and Frank Boland were building aeroplanes in 1908 with financial backing from Inglis Uppercu. *Courtesy of NJ Aviation Hall of Fame.*

Frank E. Boland (1880-1913)

Frank Edward Boland was born about 1880. Boland formed the Boland Aeroplane Company (aka Boland Aviation Company) about 1908 with his brother in Rahway, New Jersey. Boland designed, built and flew a biplane that had no tail, no rudder, and no ailerons. The pilot steered the plane by controlling two pivoting vertical surfaces at the tips of both wings. Boland's unique lateral stability system was designed to avoid paying royalties covered by patents belonging to the Wright brothers. His design attracted Inglis M. Uppercu, who financed some of Boland's experiments as early as 1908

Boland's tailless aeroplane. *Courtesy of NJ Aviation Hall of Fame.*

Boland died flying his tailless aeroplane in a crash on January 23, 1913, in Port of Spain, Trinidad, British West Indies. Fausto Rodriguez had hired the Boland Aviation Company for an aerial exhibition. Weather conditions were fine for the flight but on his attempted landing, Boland's biplane suddenly dove and struck the ground. Boland was the first to fly in Venezuela, and the first to fly in Trinidad. Rodriguez shipped Boland's body and his wrecked plane back to his family in Rahway. After his death, Boland's widow sold control of his company to Inglis M. Uppercu. In March 1914, Uppercu renamed the company Aeromarine Plane and Motor Company.

Ernest L. Janney (1893-1941)

Ernest Lloyd Janney was born on June 16, 1893, in Galt, Ontario, Canada, the son of Elizabeth and William Janney. His father was a saw maker. By 1914, Captain Ernest L. Janney was named provisional commander of the Canadian Aviation Corps. He wrote an article entitled "Aeroplanes Invaluable in European War" in 1915. Janney wrote that "The aeroplane at the front has been a huge success," and that "in one hour, the aviator can accomplish what it would take 1,000 cavalry men to do."

Captain Janney moved to Keyport, New Jersey, by January 1916 when Ensign Jason Homer Stover, head of the Aviation Corps of the Naval Reserve in New Jersey, hired him to manage the Aeromarine facilities. Stover was from Trenton, and had studied chemistry at Rutgers. In December 1915, Commander Edward McPeters appointed Stover to organize the aeronautical section of the naval reserve. Janney was on furlough from the Royal Flying Corps of Canada. By April, Janney was busy supervising the work being done on the hangar on the Aviation Training Grounds at the Robbins Brickyard tract in Keyport. Members of the Corps from the Naval Reserve of New Jersey worked quickly and work on the hangar progressed rapidly. Workers completed construction of the hangar on the property of Aeromarine Plane and Motor Company before May.

On May 29, 1916, Janney filed a trade certificate with County Clerk Joe McDermott in Monmouth County, New Jersey, to register the Janney Aircraft Company name. Its purpose was for designing, manufacturing, exhibiting and selling aeroplanes and conducting a school for flying. At the time Janney lived on Florence Avenue in Keyport. Janney's company leased an old carriage factory at Keyport from Daniel E. Mahoney. The factory had been operated for many years by Theodore Aumack. Janney planned to use nearby Raritan Bay as the testing ground for trial flights.

At the same time, the Aeromarine Plane and Motor Company donated an aeroplane to the Aviation Corp. The Naval Reserve

Flight Training School was located on the company's property. Throughout the summer of 1916, Captain Janney taught members of the Naval Reserve how to handle the aeroplane.

Safe and Sane Fourth of July

On June 21, 1916, Atlantic Highlands planned a Fourth of July celebration. Janney agreed to exhibit the plane and offer aeroplane trips for fifteen dollars. Walter and Arthur Mickens, Herbert Leonard, Herbert Hazleton and Harry B. Childs were the first passengers to book a flight. Childs, of Sea Girt, was the advertising manager for *Aerial Age Weekly* and *Flying* aviation magazines. Monmouth Country Club agreed to allow Janney to use their golf grounds as a starting and landing field. Unfortunately, things didn't go as planned.

The day before the celebration, Janney lifted off for a short flight across the meadows of Middletown and back. Upon his return, a crowd had gathered on the field blocking his landing area. Janney was forced to fly his machine into a gully in order to keep from hitting the crowd of people. Both wings were smashed. Janney was supposed to make flights at Atlantic Highlands the next day for the Fourth of July celebration but his machine was not in working order. Later, Janney talked about his aviation career and flying accidents.

> "Although I have taken up a thousand passengers, I have never caused a single one to be injured. I have been flying since 1910, and some days have taken up as many as eighty. Twice I have cracked my machines but saved the passengers from all harm. Both these times were at night in making landings under the most difficult circumstances. A storm overtook me one night when I was traveling with a passenger and I could only see now and then when lightening flashed. Finally, it struck a barn right beneath me and I made a landing by the light of the burning barn.
> I bumped into a large hog one night on a landing field in Ohio. The weight of the hog twisted my plane around so that it fell into a ditch. Landing at night on another occasion, I bumped into a fence which had just been built on the field and which I did not know about.

This was a forced landing. Otherwise I have had no accidents whatsoever, although I have had some escapes which you would might consider narrow."

By 1917, Janney moved his company to Monroe, Michigan. There he was building biplanes, and invited American and foreign military officers to inspect his new model ELJ-5 training tractor biplane, equipped with a Hall-Scott A-5 six-cylinder engine.

Ten years later, Janney was still flying. While proposing to enter a transatlantic aeroplane race of twelve aeroplanes, Janney said with a wry sense of humor:

"I hope to go in a three-engine monoplane. At the start, with a heavy load, I will have to run all the engines together. After a third of the distance has been passed over I would not need all three engines. So if one went dead I would still be able to complete the flight. I will take a radio man. The time of the flight will be from twenty-eight to thirty-five hours. If we get there, fine. If not, one more for Davey Jones' locker. What's the difference?"

One more? Did he forget about his radio man?

Janney had scheduled his flight to take off from Ottawa on July 2, 1927, and land in London, but postponed it because of difficulty selecting an airplane. Janney died on April 22, 1941, in Winnipeg, Manitoba, Canada.

Inglis M. Uppercu (1877-1944)

Inglis Moore Uppercu was born on September 17, 1877, in Evanston, Illinois, and died April 7, 1944, in New York City. He was an early member of the Aero Club of America and President of Cadillac Motor Car Company of New York. Uppercu became interested in Boland's experiments and invested in the Boland Aeroplane and Motor Company in 1908. After Boland died, Uppercu became president and chairman and in 1914 he renamed it the Aeromarine Plane and Motor Company. At that time Aeromarine was located in Nutley, New Jersey.

Unglis A. Uppercu, resident of Deal, founder of Aeromarine.
From the collection of Daniel Kusrow.

In March 1916, Uppercu relocated his company to Keyport in the northern part of Monmouth County. His wife enjoyed visiting the Jersey Shore and had bought property in Deal. He purchased a sixty-acre tract in Keyport from Cornelius Ackerson and commuted from his summer estate on Cliff Walk in Deal. Uppercu's property in Keyport was known as the Old Robbins Brickyard. It fronted on

Raritan Bay and was an ideal location for test flights. Once established in Keyport, Uppercu allowed Battalion A of the Aviation Corps of the New Jersey Naval Corps to use his facilities as a training school. He donated a hydroplane to the Naval Reserve for their use. The plane was a pusher type biplane with a six-cylinder vertical motor engine. In a pusher configuration, the propeller is located behind the motor, thereby "pushing" the plane forward.

Aeromarine Plane and Motor Company factory buildings in Keyport, July 1919. *From the collection of Daniel Kusrow.*

Aeromarine hull shop, July 1919. *From the collection of Daniel Kusrow.*

Aeromarine Engineering Department, 1922 *From the collection of Daniel Kusrow*
From left, kneeling, Kenneth Davis Vosler, born May 12, 1896, in Rural Grove, New York; US Navy 1918-1919 1st Co., NY Battalion, US Coast Guard warrant gunner; Aeromarine Chief draftsman and experimental shop manager, 1919-1924; awarded US Patent No. 1,825,632 for a percussive tool with Korvin; plant manager at EDO in Flushing NY, died August 19, 1972, in California.

Harlan Davey Fowler, born June 18, 1895, in Sacramento, California; engineer at Aeromarine 1922-1925; awarded US Patent No. 1,392,005 for a variable area wing; died April 27, 1982 in California.

Boris Viacheslav Korvin-Kroukovsky, (1895-1988); former lieutenant aviator in the Russian Army; Keyport resident and Aeromarine engineer in charge of airplane design 1921-1924; awarded many patents, including US Patent No. 1,552,259 for a retractable landing gear and another for a pontoon.

Satchy J. I.

William Kelby Warden, born June 22, 1893, in Red Bank, New Jersey; US Army WWI veteran, Mechanical Draftsman at Aeromarine; son of John J. Warden, lived at 17 Washington Street, Red Bank; retired in 1958 from Ft. Monmouth; died March 19, 1970 in Middletown, New Jersey.

Chester Learoy Horrocks, born February 18, 1894, in Marshalltown, Iowa; draftsman at Thomson-Morse Aircraft Company in Ithaca, New York; left Monmouth County for Santa Ana, California in 1923 with wife Alvira; died September 24, 1974, in San Bernardino, California.

From left, standing, Anton Bochan, born 1893, in Zamostie, Russia.

Karl H. White, born August 25, 1897, in Winchester, Massachusetts; wrote "Effect of an Airplane Rudder in Flight," *Aviation & Aeronautical Engineering,* Vol. VII, No. 7, May 1, 1918; BS in Mechanical Engineering, University of Kansas 1921; Engineering Constructor of Martin Bomber Aeroplanes; Aeromarine project engineer and assistant superintendent at Aeromarine 1921-1924, lived at 162 Church Street, Keyport; died August, 1987, in Northfield, Vermont.

Herbert W. Doherty, born 1892, in Ireland; naturalized citizen 1917; boarded at 119 1st Street, Keyport; Aeromarine draftsman 1920.

Edwin L. Davis, born 1902, in Chesterfield, New Hampshire; Aeromarine mechanical engineer; died February 8, 1967, in Red Bank, New Jersey.

Vanderlip, N.F.

Richard Young, born 1874, in Canada; lived in Matawan 1920, guard at Aeromarine plant.

Aeromarine made the U-8-D aviation motor at Keyport in 1922. They tested the 8-cylinder 180-horse power engine for overload endurance on the Dynamometer to develop reliability and durability. Capable of 150mph, Aeromarine used it in their Model 60 Passenger Limousine flying boats. *From the collection of Daniel Kusrow.*

Uppercu took an aggressive approach to his business in Keyport. The Navy awarded him a contract to build fifty Model 39-A and one hundred and fifty Model 39-B aeroplanes. These models from

Aeromarine were two-seater training planes that could be fitted with wheels or floats. In August 1917, Uppercu ordered extensive additions and new buildings built at his plant, nearly doubling its capacity. He added 30,000 square feet of working space and built a separate building to house the engine department. With the expansion completed and Navy contract in hand, Aeromarine was able to deliver between four and six hydroplanes per week. Aeromarine employed 300 workers, mostly skilled mechanics, draftsmen, woodworkers and assemblers.

Uppercu is credited with filing more than a dozen patents, including a patent for an "aeroplane" on February 5, 1919. The US Patent Office granted him Patent No. 1,367,218 on February 1, 1921. Uppercu's design featured interlocking parts for a hydro-biplane. He appointed Charles F. Redden as president in 1920.

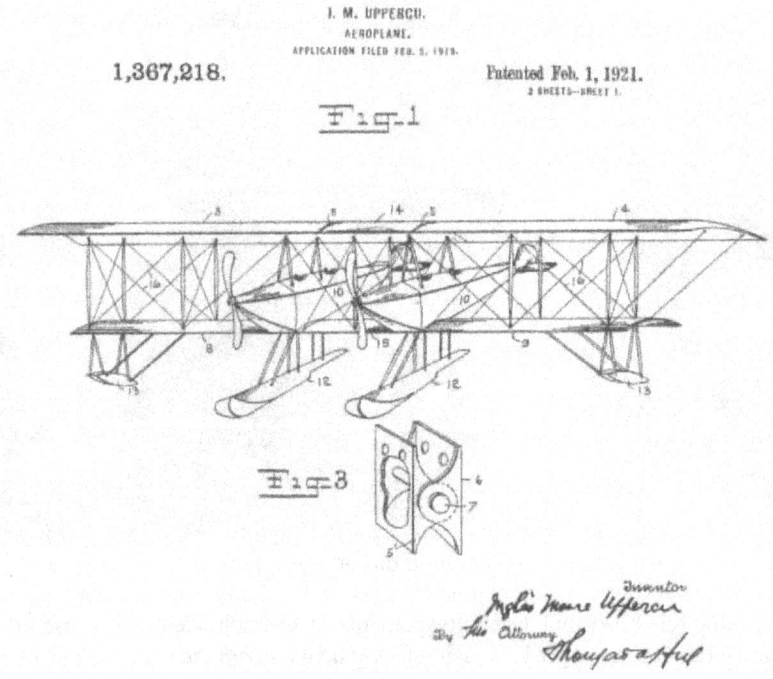

Uppercu filed for a patent on an "aeroplane" in 1919. US Patent No. 1,367,218, February 1, 1921. *US Patent Office.*

As the owner of Aeromarine Plane and Motor Company, Uppercu filed an application for a patent for a "flying-boat hull" on November 16, 1921. The US Patent Office issued him Patent No. 1,537,973, while he lived in Deal, New Jersey. Uppercu's new hull design provided for quick takeoffs and smoother, cushioned landings. Its deep V-bottom shape allowed a steeper angle of lift and descent. The new hulls were structurally sound and aided in the stability of the aircraft.

Silent film star Ethel Barrymore posed with Aeromarine's Model 50B aeroplane. *Courtesy of Keyport Historical Society and Daniel Kusrow.*

Paul G. Zimmerman (1890-1962)

Paul Gerhard Zimmerman was born in 1890 in South Dakota, and was a graduate of Rensselaer Polytechnic Institute. Zimmerman had worked for Glenn Curtiss and had experience designing and building the JN4s. He became Chief Engineer at Aeromarine in 1918. Zimmerman filed for his earliest patent at Keyport in 1910 for an "airship driving mechanism." At Aeromarine, Zimmerman designed the first aircraft with a metal hull.

Paul G. Zimmerman, Aeromarine Chief Engineer. *From the collection of Daniel Kusrow.*

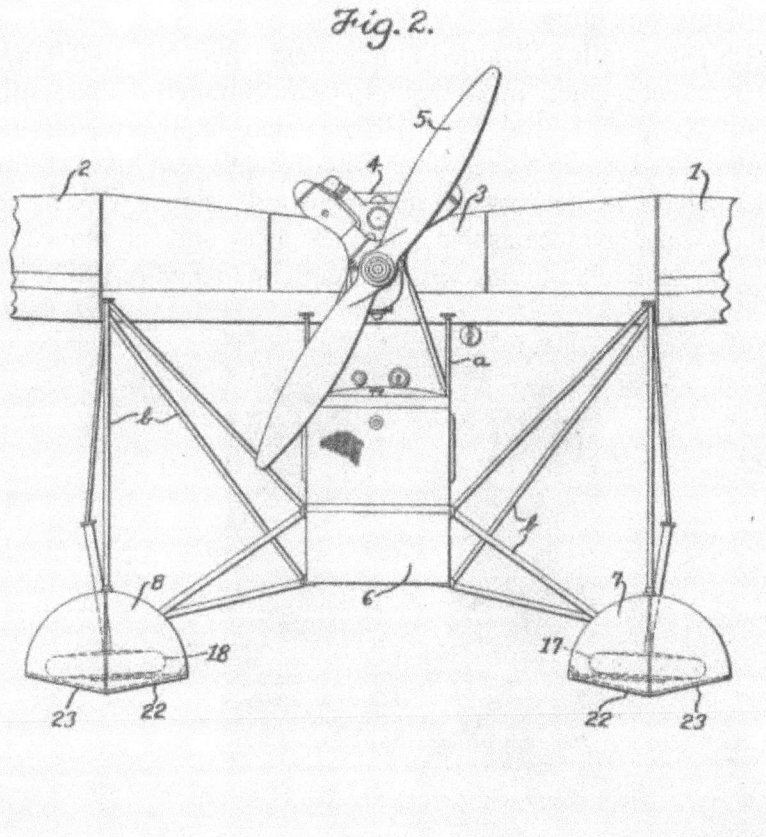

Keyport resident Zimmerman was awarded patents for drive mechanism, controlling device and landing gear. Zimmerman's "hydroaeroplane" drawing, US Patent No. 1,603,304, March 11, 1926. *US Patent Office.*

As Chief Engineer, Zimmerman ran a help-wanted ad for aeroplane layout and design draftsmen, no experience necessary. With business booming at Aeromarine, local businesses enjoyed increased profits from feeding, clothing, housing, and entertaining the newly employed workers. The company had a policy of manufacturing all the parts needed to make aeroplanes. It made its own wire, wings, motors, pontoons, propellers, canvas covers, nickel plating and more.

The US Patent Office awarded Zimmerman Patent No. 1,603,304, for a "hydroaeroplane," on March 11, 1926. Zimmerman was still living in Keyport at the time. He invented improvements for hydroplanes to land on a deck or on land. He designed his craft for the pilot to have greater forward and downward vision, a wide range to land on water and wheels that could tuck away or be employed to land on the deck of a moving vessel. Zimmerman died in 1962.

Hugh Robinson, Aeromarine Superintendent.

Hugh A. Robinson (1881-1963)

Hugh Armstrong Robinson was born in 1881 in Missouri. In 1910, Robinson was a pilot for the Exhibition Team for Glenn Curtiss. He was an inventor, daredevil and patentee. He flew at more than a dozen state fairs in the Midwest. In March 1912, Robinson attended the National Hydroplane Meet in France where he successfully demonstrated alighting and landing on rough water. Robinson designed, built, and flew aeroplanes during the very early years of aviation.

Hugh A. Robinson was the first American to make several successful flights in the Curtiss hydroaeroplane near Juan Les Pins, Nice, France, until a fast descent and high seas forced this crash. Luckily Robinson was unhurt. *Library of Congress.*

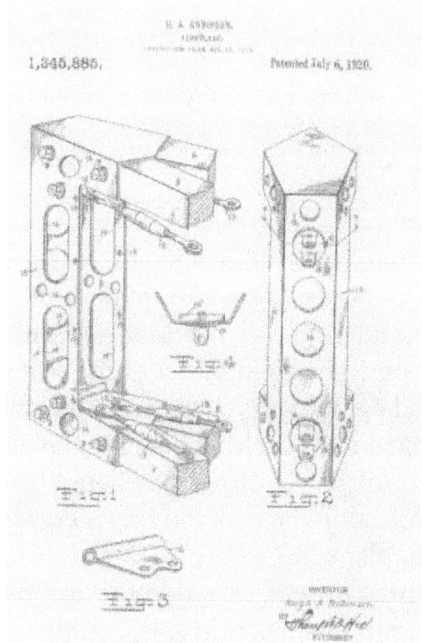

Keyport resident Robinson patented an "aeroplane," pumps and fittings. Patent No. 1,345,885, July 6, 1920. *US Patent Office.*

Robinson joined Aeromarine in Keyport in 1917 as General Manager. By 1919, he was back with Curtiss. Robinson traveled to Russia to demonstrate how useful aeroplanes could be during war. He survived several crashes and lived a full life until 1963.

Clarence A. DeGiers (1888-1987)

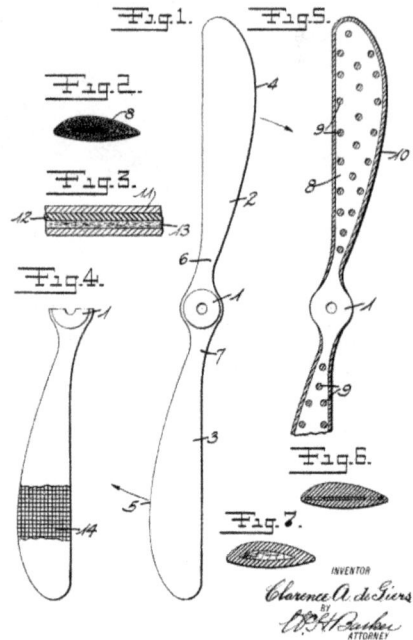

C.A. DeGiers, of Keyport, drawing for "airplane screw," Patent No. 1,384,308, July 12, 1921. *US Patent Office.*

Clarence Adair DeGiers was born in Marshall, Michigan, on June 27, 1888. In 1912, he traveled to Central America in pursuit of a $3,000 reward for the first successful flight in Panama. DeGiers followed two failed attempts, one by Robert Collier, and the other by Jessie Seligman. When he arrived, Seligman's Bleriot monoplane was still in Cristobal. DeGiers shipped it to a field at Juan Franco on the Pacific coast of Panama. On April 21, 1912, DeGiers flew successfully and won the prize. While aloft, he took moving pictures of the Panama Canal from the monoplane.

In 1917, DeGiers lived on Broad Street, Keyport, and was employed as an aviator at Aeromarine Plane and Motor Company. While in Keyport, he filed an application for a patent on an "aeroplane screw." William B. Hill and Edith J. Remond signed his application as witnesses. He assigned the patent to patent attorney William Hendry Barker, of New Brighton, New York. It took four years, but the US Patent Office finally awarded him Patent No. 1,384,308 in 1921. When business production at Aeromarine slowed, DeGiers moved to New York in 1922. DeGiers was an inventor credited with more than a dozen patents during his lifetime. He died on April 15, 1987, at Glen Head, Nassau, New York.

Roland Chilton (1890-1972)

Roland Chilton was born in 1890 in England. He joined Aeromarine as an engineer in 1917. During his thirteen years at Aeromarine, Chilton secured several patents. He amassed over 150 patents in his lifetime.

Chilton joined the Red Bank Aero Club and earned his pilot's license under instruction from Jack Casey through Airview Services, Inc. Chilton's wife Emmie also got her license there. In 1928, Chilton purchased a Waco GXE from Casey for $3,196. The Advance Aircraft Company built Chilton's Waco in Troy, Ohio. It came equipped with an OX-5 Curtiss 90 horsepower engine.

In 1930, Chilton went to work at Wright Aeronautical Corporation in Paterson, New Jersey. Because of a disagreement with Aeromarine over patent ownership and royalties, Chilton asked for a written contract upon his hiring at Wright. He agreed to assign the rights of any invention and patent to Wright, in exchange for a 2½ percent royalty on retail price. The Wright Corporation paid him his royalties but did not deduct any taxes. In his accounting method of reporting his royalty payments, the Internal Revenue Service investigated Chilton's tax returns. In Chilton vs. Commissioner, the Tax Court decided in Chilton's favor and ruled that he be permitted to treat the income as long-term gains, not part of his regular salary. He died in 1972 in Glen Rock, New Jersey.

Aeromarine Post WWI

After World War I, the demand for military planes eased. Uppercu reshaped his company into an airline passenger service. During Prohibition, Uppercu scheduled regular flights to countries where liquor flowed freely. Aeromarine booked wealthy travelers from Key West and Miami, to Havana and Bimini, and other southern and

Midwest locations. It was one of the first companies to offer regularly scheduled flights.

Aeromarine ferried passengers on a forty-five minute flight from Miami to Bimini, where alcohol was still legal during Prohibition in 1920. It was known as the "Highball Express." *Library of Congress.*

In 1928, the Klemm Company, a Delaware corporation, purchased the Aeromarine plant in Keyport. At the time, Aeromarine owned 230 acres, with much frontage on Raritan Bay, and twenty-one buildings. By 1930, suffering from financial difficulties, the Aeromarine Company was in receivership.

6. Casey Brothers

Casey's plane parked on the ice at Irwin's Boat Dock. *Courtesy of Kathy Dorn Severini DBA Dorn's Classic Images.*

James H. Casey, Jr., and his younger brother John F. Casey were the sons of Irish immigrants James Casey, Sr., and Mary Bailey, of County Cork, Ireland. Both boys were born in Shrewsbury, Monmouth County, where their father, James, Sr., worked as a farmer. The Casey boys attended public schools in Shrewsbury and Red Bank. They may have developed an interest in flying from witnessing activities of local aviators' attempts at flight. Their aviation training began with their military service during WWI.

James H. Casey, Jr. (1887-1961)

James Henry Casey, Jr., was born on July 10, 1887. On September 9, 1910, he received his US Passport No. 37,968 while he was living on Ocean Avenue in Brooklyn, New York, where he worked as a chauffeur. He was still single and living in Brooklyn when he filled out his WWI draft card registration on June 5, 1917. On December

15, 1917, Casey reported to Camp Dix in Ocean County. He served as a motor mechanic at Camp Greene, Charlotte, North Carolina, and was attached to the Aviation Branch of the US Army. By July 1918, James was stationed overseas with the Signal Corps. By October 1918, James was promoted to Sergeant, with Eighth Co., Third Motor Mechanic, Air Service in France. James mustered out of the Army between June 18, 1919, and July 25, 1919.

Casey's plane ready for takeoff on the frozen Shrewsbury River. *Courtesy of Kathy Dorn Severini DBA Dorn's Classic Images*

After his discharge James moved home. In 1920, he lived on Sycamore Avenue in Shrewsbury with his parents and siblings. He listed his occupation as an aviator. After a terrible aviation accident in 1922 James Casey moved to California. He became a police officer with the San Francisco Police Department. James H. Casey, Jr., died January 28, 1961, in San Bruno, California, and is buried in the Golden Gate Cemetery.

John (Jack) F. Casey (1894-1964)

John Francis Casey, the younger brother, was born on August 4, 1894, in Shrewsbury, Monmouth County, New Jersey. In 1910, he worked as a farm laborer on Rumson Road in South Shrewsbury and lived at home with his father James H. Casey, mother Mary and sisters Nelly and Mary.

In February 1918, John Casey enlisted in the Army, attached to the Air Service, Signal Corps, and was sent to Ground School in Princeton, New Jersey. He finished the three-month training course

Jack Casey's biplane with 1920 Stutz Bearcat. *Courtesy of Kathy Dorn Severini DBA Dorn's Classic Images.*

in technical aviation at Princeton and was assigned to the Army Flying School in Texas. Later John Casey was assigned to Rockwell Field in San Diego, California, as a flight instructor.

According to a telegram received in October 1918 by his father at home in Shrewsbury, John was commissioned as a lieutenant at March Field in Riverside, California. John mustered out of the Army between June 18, 1919, and July 25, 1919.

By 1930, John lived on Riverside Avenue in Red Bank with his wife Madeline, and worked at the Red Bank Airport as manager and aviator. John Casey, or Jack, as many knew him by, did not have children. He returned from his wedding trip in Bermuda, arriving at Ellis Island on January 9, 1930, with his new wife Madeline. In 1939, the Caseys moved to Lynn Lane, Tulsa, Oklahoma, where Jack worked as a flight instructor at the USA Aviation School. Lt. John F. Casey (aka Jack) died at his home in Passaic, New Jersey, on May 28, 1964.

The Fun Begins

The Casey brothers were discharged in the summer of 1919. Within days of coming home, they bought an aeroplane from the government at Camp Mills in Mineola, New York, where the Curtiss Aircraft plant was located. A reporter from the *Red Bank Register* incorrectly wrote that the Casey brothers were the first residents of Monmouth County to buy an aeroplane. In August 1910, John Duffy of Brielle took delivery of an aeroplane he bought and was ready to schedule a test flight; Robert Collier in Wickatunk had bought several planes starting in 1911 and in 1916, Howard S. Borden of Rumson bought a Burgess-Dunne hydroaeroplane.

John and James Casey flew their plane home to Red Bank on Saturday, July 26, 1919. Local residents came out and cheered as they brought the plane in low and waved to the locals. The Caseys landed their new plane in Aaron Armstrong's field. Armstrong, a fruit farmer on River Road, was their neighbor. The boys explained to their parents that the purpose of their investment in an aeroplane was for pleasure and for exhibition. They then hopped back aboard and flew their plane to Camp Vail and parked it in a hangar.

The Casey brothers quickly arranged for permission from the Boettner Estate on Riverside Drive in Red Bank to use the field for taekoffs and landings. They began offering rides for five dollars each, with hopes of getting a contract to sell Curtiss built aeroplanes. Timothy Quigley of Red Bank and Arthur G. Sickles of Fair Haven were the first to book their flights. Everyone who got the chance to fly was enthused. In the first two weeks, the Casey boys flew many passengers.

Not All Strictly Business

Lars (Larry) Olsen lived on Riverside Drive and worked at the Aeromarine Plane & Motor Company in Keyport. On weekends, Olsen helped the Casey brothers with their weekend flights. At his regular job at the airplane plant in Keyport, Olsen worked with Middletown Township Committeeman John M. Johnson. One day, Olsen and Johnson were discussing the usefulness of aeroplanes. The two disagreed. Johnson thought they were only good for thrills and pleasure. Olsen believed the aeroplanes could be used for more purposeful business endeavors, such as delivering mail. In those days "air mail" was a rudimentary method of bags of mail being picked up at one place and "dropped" at another. Johnson said pilots would not be able to insure that the bags of mail would land in the proper place. Olsen, who had some flying experience with the Casey pilots, disagreed, and the bet was on. Johnson offered a ten-year-old bottle of applejack if Olsen could hit the rooftop of Johnson's house in Belford with a cabbage. If Olsen could do that, Johnson would hand over his aged choice bottle of whiskey and never again question the usefulness of an aeroplane being used to deliver mail. What Olsen wagered is unknown. Johnson was so sure of winning his bet, he bragged about it at the Belford Post Office. Word spread about town that Olsen and the Casey boys were going to give it a try.

A few days later, Johnson was at home reading his evening newspaper when he heard a loud thump. He jumped from his easy chair and saw the plane circling his house. As he watched from his front yard, Olsen threw a cabbage at his house and Casey threw another one, knocking some roof shingles loose. Casey and Olsen threw a few more cabbages, another landing on the Belford post office roof. Certainly, Casey & Olsen had to be smiling on the return flight. It was a dramatic victory. Johnson's neighbor told him, "You'd better pay that bet without delay. If I were you I would not fool around with those boys. If you do they are apt to fly over your house again and throw down a few bricks!" Johnson took his neighbor's advice and handed over the bottle that night.

When Pigs Fly?

Casey and Olsen were convinced that aeroplanes were the way of the future and that flying machines could be used for practical business purposes. Uses were only limited to what they could imagine, and in this case, their imagination ran a little wild. They offered to deliver small pigs by parachute, guaranteeing that the pigs would land on the correct spot on the buyer's property. What they really wanted was a contract with the US Postal Service to deliver air mail. They practiced picking up mail pouches placed on tall poles while traveling in midflight. They advertised that they were available to drop rice at weddings on the newlyweds. Obviously obsessed with developing flight into a prosperous business, Olsen began designing plans for an aeroplane that would only cost $2,000.

A new passenger and flight enthusiast was J. Clark Conover of Holmdel, owner of the Strand Theater in Red Bank. Conover asked Casey to distribute several thousand flyers advertising *The Right To Happiness,* which was appearing in his theater. Actress Dorothy Phillips starred in this 1919 silent movie about crazed Bolsheviks. They flew over Red Bank and Long Branch and distributed the flyers.

In October, Jack Casey and J. Clark Conover flew from Red Bank to Atlantic City and back. The flight south took them one hour and fifteen minutes. The return flight took forty minutes longer due to strong head winds. Weather conditions were good both ways. In total they covered about 200 miles in two hours and fifty-five minutes without mishap. While there, they spent about four hours walking the Atlantic City boardwalk. In those days, a fast express train could only make a one-way trip from Red Bank to Atlantic City in about three hours. They proved they could do it in half the time. News of Casey and Conover's speedy, daring round trip made headlines.

Before the flight, Conover met with local undertaker Albert W. Worden. Conover gave Worden instructions on how he would like to be buried, just in case he was killed. Upon landing in Red Bank, Conover said he enjoyed every minute of his trip and that traveling by plane beat traveling by automobile in every respect. He said he felt safer in the sky with Casey as the pilot than he did in an automobile on the roads. It was enough to convince Conover to begin taking flying lessons from Casey and to buy his own flying machine.

Casey and Conover planned to travel by air to attend the Princeton vs. Harvard football game at Palmer Stadium in Princeton on November 8, 1919. On the way they lost their bearings, so they landed at Somerville to ask about their location. They continued on and landed at Princeton but got there too late. The game ended in a 10-10 tie. A black and white video of that game has survived and is available on www.youtube.com.

Back home Casey continued to offer rides for five dollars. In November 1919, the Curtiss Aircraft Company awarded Jack and James Casey a contract to sell biplanes and hydroplanes. They looked for a suitable showroom in Red Bank, and promised to have a "flying boat" on the Shrewsbury River by the next summer. The Casey brothers hoped to earn a living by selling planes to the wealthy residents of Monmouth County.

Jack Casey celebrated the 1920 New Year's Eve with a bang, with lots of bangs. Casey shot off fireworks from his plane over Red Bank, and made frequent trips to replenish his supply.

In April, the railroad labor union workers went on strike. Workers wanted to be paid more than forty cents per hour and to work less than their 10-hour per day. The strike shut down rail service and affected the entire country. Passenger service halted in Monmouth County; workers could not commute to New York by train. Mail was not being delivered. J. Clark Conover cancelled the afternoon matinee at his Strand Theater because the films didn't arrive. The Casey brothers saw the opportunity to find another use for their aeroplanes. Why not fly to New York and pick up the films? It was a good idea, but the strike ended before they made the trip.

In May 1920, Jack Casey and J. Clark Conover flew to an altitude of 11,000 feet above sea level. They rose high enough to look down onto the clouds. Spectators watched the outline of the plane shrink in size then disappear into the whiteness of the clouds. A few moments later they saw the aeroplane emerge from the cloud cover, with the plane's tail facing downward as it dove toward the earth in a spinning motion. The pilots eventually pulled the plane out of the tailspin and climbed back up into the clouds. The plane went into a second tailspin and the pilots were again able to level off. As they came out of the clouds a third time, Casey and Conover realized they were over the ocean off Asbury Park. Casey turned off the engine and glided home to the landing field on Riverside Drive. Ice from the higher altitude and colder temperature covered the plane but began to melt rapidly once on the ground.

When asked about their trip, Casey and Conover exclaimed in unison "It was great!" Plunging through the clouds was the most eerie and uncanny sensation they had ever experienced. They described the clouds as much denser than fog on land and that being in the clouds was the same as being lost. They said that after they had been in the clouds a short time, they lost all sense of direction and were unable to tell whether their machine was going up or down or whether they were riding upside down or right side up. They reported that while they were in the clouds they were unable to realize that the machine was in a tailspin. Casey and Conover said it was not until they fell out of the clouds and saw the earth revolving that they knew that their aeroplane was performing exceedingly well. Conover said:

"Red Bank, Long Branch and Asbury Park looked almost like one continuous town, the distance between them seemed so short. Yes, and when we fell through the clouds and saw Red Bank, Long Branch and Asbury Park doing a shimmy dance, we knew we were in a tailspin."

Casey added, "The rapid revolving motion made us sick at our stomach for a short time, but we soon got over that." According to a report in the *Red Bank Register,* "The men said that above the clouds there was a steady continuous gale, but no violent puffs. They said that the clouds rolled by under them at a tremendous rate and that they looked like huge ocean billows. The rays of sun shining on the wavy clouds made a beautiful picture, the aviators said, and both men regretted that they did not take a camera along with them. So far as known the flight was the highest that has ever been made with the same type of machine as is used by the Casey boys." It was a lofty flight but not a world record. In 1915, Joseph E. Carberry, one of the first military pilots, set a US Army altitude record by flying to the altitude of 11,690 feet in a Curtiss Model G biplane.

In June 1920, Casey contributed to the success of the Memorial Day Parade in Red Bank by showering down 50,000 tiny American flags as souvenirs. With his brother James, the Casey boys shared some success as agents for the Curtiss Aeroplane and Motor Company. They sold a three-passenger flying boat to Colonel Howard S. Borden of Rumson. Borden paid $12,000 for the Sea Gull type plane that featured plush upholstery and many extra deluxe conveniences.

Casey also profited by using his airplane to provide advertising to local businesses. The owner of the new Broad Street Cleaners and Dyers at 48 Broad Street hired Casey to disperse flyers over the area announcing their grand opening. Other merchants noticed and hired Casey to use this novel way of advertising.

Casey and his brother were in negotiations to lease a field on the corner of Harding Road and Ridge Road as an airfield for landings and takeoffs. The field was centrally located and seemed an ideal place to store aircraft purchased by local residents.

Early Aviator Ruth Law. *Library of Congress.*

From August to October in 1921, Casey and Conover made one of their longest trips, from Asbury Park to Omaha, Nebraska, and back. They flew 6,000 miles in two months, in a Curtiss Special, as part of Ruth Law's Flying Circus. Law had hired Casey and Conover as a backup for her three-plane aerial show. They made stops at several state and county fairs in the Midwest and amazed many spectators. Their first fair was Sedalia, Missouri. They then went onto Toledo and Cleveland, Ohio; Des Moines, Iowa; Bowling Green, Kentucky; Grand Rapids, Michigan; and Omaha, Nebraska. On their return, Casey commented that Midwestern families showed them much hospitality.

In November 1921, Casey, Conover, and a third passenger, photographer Howard Cole, flew from Red Bank to New York City and back in one hour and three minutes. They used a flying boat made by the Aeromarine and Motor Company in Keyport. Upon reaching the city, Casey flew up the East River and the Harlem River then cut across and followed the Hudson River back. The party landed on the Shrewsbury River and parked the hydroplane at Charles Irwin's dock.

Captain Charles P. Irwin, builder of motor boats and ice boats. *Courtesy of Red Bank Public Library Photograph Collection.*

Tragedy Strikes

On January 15, 1922, James Casey lost control of his Curtiss biplane while it was still on the frozen Shrewsbury River, amid ice skaters, ice boaters and ice fishermen who were enjoying the day on the ice. His plane was equipped with a double cockpit and both Caseys had been taking passengers up for rides all afternoon. Until that day, James Casey and his brother were regarded as expert pilots and had not suffered a single fatal casualty.

Casey's plane, while still on the ice, struck and killed Mrs. Anna C. Hounihan who was there with her husband Raymond, their two young children, Grace and Harold, and Anna's brother, Lawrence Conley. Conley's arm was severed below his elbow by the mahogany propeller blades as he reached out to save his sister. A score of other people escaped being slashed to death by a matter of inches. Dr. Charles G. Sherf was there on the river with others enjoying the skating. When the accident occurred, Dr. Sherf used Hounihan's torn clothing as a tourniquet to stop Conley's bleeding. Another skater nearby, Wesley Walker, was knocked down and bruised badly.

Casey had helped his next paid passenger into the rear cockpit then prepared for takeoff. He went behind the plane and checked to see that the tail brake was set to hold the plane on the ice. He set the blocks under the carriage wheels, adjusted the spark, gave it a little gas and grabbed the propeller. Casey gave it the first twist and immediately noticed something was wrong. Somehow the throttle accidently sprang wide open and the motor started with a roar. The blades were rotating at full speed. The plane swayed; it jumped its chucks and began rushing straight forward. Casey ducked under the blades and tried to grab the underbody but was unable to get a hold. He managed to grab onto the left wing tip but was not able to restrain it. Either a bump in the ice or a puff of wind put the plane on a course directly into a crowd of curious onlookers. Bystanders grabbed the right wing which turned the plane but were shaken off. The plane was occupied only by a frightened, helpless passenger in the rear cockpit, while Casey held onto the tip for dear life. The plane sped up and stayed on the ice. Eventually Casey was able to swing himself into the cockpit and turn off the engine switch. The passenger was unhurt; the propeller was smashed. Mrs. Hounihan died enroute as they rushed her to Dr. Edwin Field's office on West Front Street. Volunteers drove Conley to Monmouth Memorial Hospital in Long Branch where doctors amputated his right arm.

Red Bank Police Chief Harry H. Clayton arrested James Casey and charged him with manslaughter for the death of Anna Hounihan. Monmouth County Judge Rulif L. Lawrence released Casey in Freehold on $1,500 bail posted by Charles P. Irwin. Casey was also brought before Justice of the Peace Edward Wise in Red Bank and charged with assault and battery for the injury to Lawrence Conley. The next day, Casey explained in his own words what had happened to a reporter.

> "There was nothing to be alarmed at when the aeroplane crept forward. If the crowd had only kept hands off the machine and left matters to me, the accident would never have happened. They pulled the machine right into the crowd. Lars Olsen, my assistant, had hold of the other wing, but the combined strength of the crowd was too much for him and he couldn't keep the machine straight. The aeroplane was headed straight into the wind towards Cooper's Bridge when I started the engine. It frequently happens that an aeroplane will move forward as soon as the propeller starts revolving just as an automobile will sometimes do this. Such an occurrence is not dangerous. It has often happened to me and other aviators and it has never resulted in an accident as far as I know. If the aeroplane had been allowed to come forward it would have been easy for me to step aboard it, just as I have often done when this has happened before. But the crowd thought the plane was running away and they tried to hold it back. They couldn't have done a worse thing. The side of the aeroplane I wanted to get on kept going away from me and I had to get in on the other side, thus losing considerable time."

At the following meeting of the Red Bank Council, members decided to write an ordinance banning motorized craft on the ice because of the danger they posed. The Red Bank Knights of Columbus raised $500 for the benefit of Lawrence Conley.

The trial of State of New Jersey vs. James Casey was scheduled to be heard in the Monmouth County Quarter Sessions Court in Freehold on February 9, 1922.

Wild Bill Casey

Jack Casey taught his nephew William J. Casey how to fly at Airview Field in the late 1930s. During WWII, William Casey became known as Wild Bill Casey. Lt. Casey was part of the 306th Bombardment Group in Europe. He saved a crippled plane during a bombing raid on the Nazi troops concentrated in France. When Casey saw a wounded plane, he would fall out of formation to attract German war planes. On one such maneuver, Casey's B-17 plane was shot down on April 7, 1943. He spent the rest of the war as a prisoner in Stalag Luft 3.

In October 1924, J. Clark Conover bought his own plane from Casey. Conover had retired from the motion picture business. The wealthy retiree was accustomed to purchasing a new automobile every year but decided to switch transportation modes. He said "There is more room up there, less traffic, and less danger." Conover may have had second thoughts about how dangerous flying really was, just a week after he purchased the aeroplane. On a training flight with Casey, about 500 feet above Camp Vail, the engine exploded and the aeroplane caught on fire. Casey made a safe landing on the parade grounds but suffered burns on his face and hands.

Airview, Inc.

On June 30, 1926, Casey established Airview, Inc., with the Johnson brothers of Long Branch. George R. Johnson, vice-president of Airview and main photographer, was a veteran of the Fourteenth Photo Section during WWI and a member of the 101st Photo Section in Germany after the war. His brother Garrett I. Johnson was a Rutgers graduate and was named secretary, treasurer and engineer of Airview. Their main objective was aerial photography for advertising. With Casey flying the plane, Johnson took photos of towns, land surveys and real estate developments. The Standard Oil Company hired them to take pictures of their buildings. Owners of the Berkeley-Carteret Hotel in Asbury also hired them to take aerial photographs for advertising.

That November 1926, Casey bought a twelve-acre farm on Shrewsbury Avenue from Morgan F. Larson and his wife Margaret. Casey began converting the farm into an airport. During the spring of 1927, Casey worked hard to make the field acceptable according to regulations of the US Department of Commerce. He built a hangar and painted "RED BANK" in fifteen-foot tall letters in black against a yellow background on the roof. He installed a windsock, electric lighting, underground fuel tanks and a telephone booth. He painted a 100-foot diameter circle in the middle of the landing field. With Casey's hard work, Airview Flying Field began to see traffic the first week of May 1927. Several planes landed. The pilots thought the field was well marked and in good condition. Casey bought another adjoining twelve-acres from Harold G. Hoffman and his wife Lillian in July 1928.

Commander Richard E. Byrd. *Library of Congress.*

Red Bank Aero Club

The Red Bank Aero Club was organized on October 11, 1927, for the purpose of establishing a municipal airport in or near Red Bank. Morton V. Pach, local cigar dealer, had talked about forming an aero club with Jack Casey a year earlier. Both were eager to move things along. Members of the Aero Club elected Theodore Parsons president, Jack Casey vice-president, George R. Johnson secretary and J. Clark Conover treasurer. The thirteen original members also included Lt. Vernon Treat, William Wyckoff, William Manna, John H. Kelley, James K. Hanigman, Lloyd Sickles, Herbert Henenberg, Louis Olson, and Benjamin L. Atwater.

Slowly the idea of a local airport gained steam. Members met for the first time in the law offices of Quinn, Parsons and Doremus, Esqs. They asked members to contribute to a fund to purchase a plane, which was to be used for flying lessons. They modeled their organization's Articles and Bylaws on other already existing aero clubs around the state, including New Brunswick, Atlantic City, Camden and Newark. In March 1928, Commander Richard E. Byrd gave a lecture and helped raise $1,000 towards establishing an airport in Red Bank.

Red Bank cigar dealer, Morton V. Pach, supported the idea of a local airport. *Courtesy of Red Bank Library.*

Jack Casey quit his job in April 1928 to devote himself full time as owner of the Airview Flying Field at Red Bank. He acquired the dealership for Waco planes. In January 1929, Casey hired a company from Maryland to build an 80x80-foot hangar, then doubled the hangar size with an addition that September. The *Red Bank Register* dubbed Casey the "father of flying" for his interest and dedication to local aviation.

Brunner builds an Airplane

One of the Aero Club members, Adam Brunner of Englishtown, was the first to build an airplane and keep it in the new hangar at the Airview field. Brunner modeled his craft after the Irwin Meteor type, one of the smallest planes built. Brunner's plane was eleven feet long and less than twenty feet wide, wing tip to tip. It had a single seat, and was powered by a four-cylinder, air-cooled, 20 horsepower engine. It took Brunner about three months and cost him about $800 to build.

Airview Flying Service, Inc.

Jack Casey, Valentine Van Keuren of Eatontown, and J. Clark Conover of Red Bank, formed the Airview Flying Service, Inc., on August 2, 1928. Casey received ten preferred shares and ten common shares of stock; Van Keuren and Conover received five shares of common stock each. Airview purchased the land the airport was on from Casey in December 1928.

Holger Hoiriis. *Courtesy of Colin Hunt.*

Red Bank Airport

The Red Bank Airport was originally called the Airview Flying Field as early as May 1927. By October 1927, Casey kept his Waco plane there along with a plane owned by Lt. Vernon Treat, former aviator with the Third Division during WWI.

Airview, Inc., built the Red Bank Airport, and Airview Flying Services operated it. While he managed the airport, Casey gave lessons to Holger Hoiriis, a renowned Danish born pilot who flew from New York to Denmark in 1931.

The property that Casey purchased and used as Airview Flying Field was actually located in Shrewsbury Township. That section of Shrewsbury was named New Shrewsbury Borough in 1950, then renamed Tinton Falls Borough in 1975.

On February 13, 1929, Monmouth County Clerk Joseph McDermott issued a certificate of incorporation in the name of "The Red Bank Airport," to Mattie Van Brunt, Dorothy E. Turkington, and Marjorie Uprichard. Thomas P. Doremus, Esq., who Casey had retained in earlier legal matters, signed as legal agent for the corporation.

Turkington and Uprichard were very young women. Both had recently graduated high school and worked as stenographers in a lawyer's office. Van Brunt was a fifty-one-year old, unemployed housewife from Long Branch. Each received five shares. It is not known why these women were named stockholders of the airport on the incorporation papers.

Red Bank Races, Inc.

On March 26, 1929, County Clerk McDermott issued another certificate of incorporation in the name of "The Red Bank Air Races, Inc." to William B. Harding of Holmdel, John F. Casey, and Valentine Van Keuren.

Van Keuren, a civil engineer, later traveled to Peru on an expedition to take aerial photos as part of the Magellan Expedition. Casey's photographer, George Johnson, was also a part of the Peruvian expedition. Harding had recently earned his pilot's license at Airview. To qualify for a pilot's license in the 1920s, a student was required to attend ten hours of class instruction, pass a written test and log ten hours flying time. As usual, Jack Casey was in the middle of anything to do with flying in Red Bank.

In April 1929, the Fox News Company filmed Casey in an eight-minute movie about the upcoming aviation meet in July at Red Bank Aero Club. Fox produced it to show in theaters along with their featured movies.

The Red Bank Aero Club held the Red Bank Air Races July 3-6, 1929. They scheduled events and offered prizes for a thirty-mile race, for women pilots, and for accuracy landings. Strict rules prohibited pilots from wild flying. Aviators were not allowed to "jazz" the crowd; no unscheduled stunts from any height were permitted; no freak landings or no steep slides were allowed.

Later that year, in September, the airport bought the thirty-five acre farm formerly owned by Georgianna Patterson, bringing the airfield acreage to 200 acres.

Red Bank Airport circa 1950s. Walter R. Laudenslager, *right*, owned the airport 1938-1963. *Courtesy of Kathy Dorn Severini DBA Dorn's Classic Images.*

Casey sold his interest in the airport to Walter R. Laudenslager in 1938. Laudenslager owned and managed the airport until 1963. The Red Bank Airport closed in the 1970s, after several fatal accidents.

7. De Luxe Air Services, Inc.

Stuart A. Morgan and Franke Benjamin's official wedding photo taken at Keesler Air Force Base, Biloxi, Mississippi, c.1947. *Courtesy of Ardy Tobin.*

Stuart A. Morgan (1893-1975)

Stuart Alexander Morgan was born on September 23, 1893, in Richmond, Virginia. His father Frank was an English immigrant and worked as a private detective. Morgan served in the Aviation Section of the US Army during WWI as a flight instructor then remained in the Reserves until being activated in WWII. Still single in 1930, he worked as an aeronautics inspector for the Aeronautical Branch of the US Department of Commerce. By 1940, Morgan was married and worked as a real estate developer. He died January 5, 1975, in Ft. Lauderdale, Florida.

Lt. Col. Stuart A. Morgan. *Courtesy of Ardy Tobin.*

Pre-WWI Test Pilot

At Staten Island, New York, Morgan worked for Charles R. Wittemann. Wittemann is credited in 1905 with opening the first commercial airplane manufacturing plant in the world. Morgan was a test pilot for some of the flying machines produced at the Wittemann facility during 1914-1916.

Military Service

Morgan registered for the draft while he lived in Staten Island, New York. He entered as a candidate for commissioned officer in the Reserve Corps at Plattsburg Barracks, New York. Morgan was called into service March 25, 1918, and was assigned to the School of Military Aeronautics at Cornell University. He served as a flight instructor stationed at Kelly Field in San Antonio and Camp John Dick in Dallas, Texas; Carlstrom Field and Dorr Field near Arcadia, Florida; and Souther Field in Americus, Georgia; before being honorably discharged with the rank of Second Lieutenant on December 10, 1918.

Post-WWI in Asbury Park

Asbury Park Mayor Clarence E. Hettrick. *History of Monmouth County, New Jersey, 1922.*

After being discharged, Morgan entered into an agreement with Asbury Park Mayor Clarence E. F. Hettrick, to oversee the new aviation field at the old Monmouth County Poor Farm west of Asbury Park. On May 1, 1919, he and pilot Thomas B. Lyons began offering daily passenger flights lasting fifteen minutes or more.

Morgan started with one airplane and planned to add more as business increased.

Airplane Crashed

A month into providing airplane flights to passengers, Morgan had trouble landing with a passenger aboard. As he came in for a landing, a man ran onto the field, darting into the path of the incoming plane. Morgan acted quickly, "putting on more speed," and avoided hitting the man but was unable to slow the plane before running out of cleared field. The plane collided with a tree at the edge of the landing field. Fortunately, neither pilot nor passenger was hurt. Morgan sold the plane to Jay A. Mellish of New York, and bought a new Curtiss airplane to replace it. He took delivery at Atlantic City and flew it back to Asbury Park.

De Luxe Air Service, Inc.

On March 29, 1920, Morgan established De Luxe Air Service, Inc., along with shareholders Lt. Harold A. Steiner and local businessman Walter Steinbach. Its principal office was at 701 Mattison Avenue, Asbury Park. Steinbach of the Steinbach Company of Asbury Park was vice-president. De Luxe Treasurer was Lt. Steiner, a Yale graduate who owned a pajama manufacturing company. Steiner enlisted for military service on July 13, 1917, and graudated the Princeton School of Military Aeronautics in October 1917. He spent time overseas flying with the 184th Aero Squadron. Steiner listed his residence at the Metropolitan Hotel in Asbury Park.

Morgan established De Luxe Air Service to buy, lease or sell land and buildings for establishing airplane fields and a flight school, for building aeroplanes, aerial surveying, carrying passengers, aerial photography and aerial advertising. Through De Luxe, Morgan leased airfields in Lake Como and in West Deal near the Hollywood Golf Course, where they offered gasoline, supplies and repair services to cross country pilots. They built a hangar large enough to house four Curtiss aeroplanes. Morgan added regular markings to the field so pilots flying over could identify it as a landing site. The company stocked parts for Curtiss planes and OX Motors and were distributors for Titanine, a company in Union, New Jersey, which made dopes and covering materials. De Luxe advertised large supply of parts, prompt shipment and fair prices.

By December 1920, Morgan was not associated with this company and Charles D. Winters was president. Winters hired former Army aviators Lt. Ted Parsons, Lt. Frank Turner and Lt. Arthur McAleenan, Jr. Winters had a policy of only hiring WWI veteran aviators. Winters planned to operate De Luxe as a seasonal company, offering airplane rides starting up with good weather in the spring and closing down in the fall.

Theodore D. Parsons (1894-1978)

Ted Parsons was born in 1894 in Wisconsin. In 1917, Parsons worked as a law clerk Red Bank. He served during WWI in the US Army Air Service as a test pilot in the 139th Aero Squadron.

After his discharge Parsons passed the bar in 1919 and began practicing law in Asbury Park. He worked as a pilot for De Luxe in the early 1920s. In 1929, the Red Bank Aero Club formed and members elected Parsons as the club's first president. Parsons was later named New Jersey Attorney General.

Lt. Vernon E. Treat (1894-1978)

Vernon Egbert Treat was born on April 6, 1894, in Fresno, California. He was an aviator during WWI in the American Expeditionary Forces (A.E.F.) and later worked as a stunt pilot as part of Ruth Law's Flying Circus barnstormer exhibition performances at state and county fairs.

Lt. Treat was named head pilot of De Luxe Air Service. In December 1920, Treat flew from Asbury Park to Birmingham, Alabama, to practice aerial performances. Ruth Law had hired him through De Luxe for a barnstormimg exhibition schedule through the midwest which was to start in May. Along the way, Treat gave exhibitions and carried many passengers. Treat used city streets for landing and takeoffs. He returned in January 1921 without incident, reporting that he needed more power, greater stability, faster climbs, slower landing speed, and a greater factor of safety. In March 1921, Frank Turner took over as president of De Luxe. In May, Treat flew to Detroit to work in Ruth Law's Flying Circus, logging another

1,500 miles. As a backup aircraft, De Luxe sent Jack Casey in a five-passenger Curtiss airplane.

A WWI Veteran and Wall Township resident, Lt. Vernon Egbert Treat later became an air mail pilot. *Rich701 at www.Flickr.com.*

Casey had operated a flying service in Red Bank for two years but was struggling to make ends meet. In August 1921, Casey, along with J. Clark Conover, flew to Sedalia, Missouri, to join Treat and Law, and arrived just in time. Treat was having engine trouble, which he blamed on some bad oil he took on at a service station. Casey made the trip from Asbury to Sedalia in three days. From Sedalia, Treat and Casey flew to Omaha, Nebraska, then Battle Creek, Michigan, and back to Trenton, New Jersey. They logged another 6,000 air miles, carrying passengers and performing daredevil stunts, including wing-walking exhibitions and night flights with fire works, all without incident. Finally back in Asbury Park, they logged over 10,000 cross country miles, and 300 hours. The only damage was to two propellers, which broke when they tried to land in muddy, bumpy fields. Not one passenger suffered as much as a scratch. De Luxe claimed "an aeroplane is as safe as its pilot." In review of 1921 operations, Turner claimed De Luxe Air Service set a record for longest distance flying a Candian Curtiss airplane without incident.

After his stints with De Luxe and Ruth Law, Treat lived in Wall Township and worked as an Air Mail Pilot. In 1940, he was a flight instructor in Howell Township. In 1942, he was in Tulsa, Oklahoma, employed with the Spartan Aircraft Company. He worked as an Airman Standards Inspector for the US Department of Commerce, Civil Aeronautics Division in Miami, Florida, in 1949. Treat died on January 25, 1978, in Greenville, South Carolina.

A Wedding in the Clouds

In keeping with creative ideas to keep the company afloat, De Luxe offered to supply for free an airplane, pilot and minister, to the first couple willing to get married aloft. Other flying companies offered the same package so it wasn't unique, just a bit odd, to get married in an open cockpit aeroplane. It is not known whether any couple took them up on their offer.

In January 1922, the Asbury Park Chamber of Commerce sponsored the Second Annual Asbury Park's Businessmen's Show. It was a week-long celebration of "free, joyous entertainment." Over 100 businesses exhibited, and De Luxe Air Service was a big part of it. De Luxe featured "The Loops of Death, Outlined in Red Fire." De Luxe advertised it as the most thrilling night-flying spectacle ever devised. At 8:00 P.M. on January 23, 1922, Lieutenant Verne Treat was scheduled to fly about 2,000 feet above the Asbury Park Casino and Fishing Pier, and perform a six-loop, nonstop flight pattern.

Treat thrilled the crowd as planned. He flew in from Sea Girt. High above the jam-packed Asbury Park Boardwalk and Casino, Treat performed his loops of death stunt, with red lights attached to the wings. His plane gave off "a stream of colorful fire."

Arthur McAleenan, Jr. (1894-1920)

Arthur McAleenan was born in New York City in 1894. He graduated from Yale University in 1915 and was a diving champion.

McAleenan enlisted in the Army in May 1917. He trained at the US School of Military Aeronautics at Cornell University and became a flight instructor at Ellington Field, Texas. McAleenan was discharged as a First Lieutenant on January 5, 1920. He moved to Deal, New Jersey, and signed up as a pilot with De Luxe Air Services, but died a few months later as a result of a car accident.

8. Monmouth Builders and Flyers

Ralph L. Bray (1872-1945+)

Ralph Bray from Long Branch was an early Monmouth County aviator, circa 1909. *Aeronautics, June 1911,*

Ralph Limon Bray was born on May 1, 1872, in Long Branch, New Jersey, the son of Joseph & Harriet Bray. His father was a carpenter. In 1896, Ralph married Nettie Northam in the Great Auditorium in Ocean Grove. From an early age, Bray was an inventor. The US Patent Office awarded Bray Patent No. 880,714A for a laundry-edger on March 3, 1908. Bray was a Mason of the third degree, a member of the Red Men and Golden Eagles.

From 1909-1912, Bray was selling pianos, New Home sewing machines, and 7 horsepower M.&M. motorcycles out of his store at 560 Broadway in Long Branch. He lived at 69 Grand Avenue and dealt mainly in musical goods. Bray took his family to Florida via automobile for winter vacations. In 1916, 1920, and 1921, he drove to St. Petersburg. In 1922, Bray ran a classified ad in the *New York Times* to sell his piano business due to poor health. His business had been established for twenty-three years. Bray asked $4500 for his inventory of forty pianos, a delivery truck, and good will.

Bray also maintained a winter residence in Cold Spring Village of Philipstown in Putnam County, New York. In 1930, Ralph, Nettie, their married daughter, Gertrude, and their son-in-law, William A. La Due lived there, all unemployed. Ten years later they all were living at 695 NE70th Street in Miami, Dade County, Florida, where Ralph remained until at least 1945. His whereabouts are unknown after that.

Bray's 2-cycle engine. *Smithsonian Institute, National Air & Space Museum (NASM 81-15187).*

Bray's Flying Machine No. 1

Ralph Bray was one of the earliest Monmouth County residents to try his hand at flying. Bray was building aeroplanes of his own design before 1910 in Long Branch. It was first reported in November 1909 that Bray was building an aeroplane modeled after the Latham type, and that he expected to fly before the end of that year. An article in *Science,* July 23, 1909, stated that the design of Hubert Latham's monoplane would rival the Wright brothers' patented machines, and that it had already been proven to be faster. Latham called his machine *Antoinette.*

Bray's monoplane was powered by a two-cylinder, 30 horsepower Detroit Aero motor, which turned a seven-foot propeller and could develop a thrust of 180 pounds. The motor weighed 120 pounds and cost $250 dollars, which was inexpensive compared to other similar aircraft engines available at the time. Bray positioned the engine low enough to avoid having to cut a notch in the upper wing.

Bray fashioned the framework of his machine out of bamboo poles between 1½ inches thick and used sheet copper to make the connecting joints. The wings on his experimental craft spread twenty-five feet from tip to tip and required 379 square feet of canvas to cover them. Bray worked out of a basement in Odd Fellow's Hall, to design and build his aeroplane.

Bray once told a reporter that on one occasion, he opened the throttle a little too far and his plane got off the ground before he was aware. In trying to dodge a flock of crows, he turned too short and caught the end of his wing, causing a little damage, which he repaired locally.

Stuck in the Mud

Bray continued working on his monoplane for the first half of 1910 until the first test run on June 18, 1910. He carted the plane to a field near the Shrewsbury River and faced it into the wind. These early test flights always attracted a crowd. Bray extended the wings and positioned himself at the steering controls. His assistant, Willis Wood, gave the propeller a turn and the engine started. Willis pushed the machine mounted on three bicycle wheels along the road for some distance but the machine never got airborne. Bray's machine got bogged down in the muddy field. He tried to make a second attempt going with the wind but had no success. A reporter for the *Ocean Grove Times* wrote that "one of the wings became inflated and would have turned the flyer over had not assistants come to the rescue." Bray postponed more trials until the next week. Eventually he scrapped his original design and decided to build a second one.

Flying Machine No. 2

Less than thirty days later, in July 1910 Bray was working on his second flying machine. This one was a two-winged flying machine modeled after the Curtiss biplane. It took him three months, but by September 1910 he was waiting for the engine to arrive and was ready to test it. Setbacks stalled the testing of his second plane for another eight months. Bray stored it in the YMCA building on Branchport Avenue in Long Branch while he waited for an engine to arrive.

The upper wing on Bray's second plane spanned thirty feet across. The lower wing was just twenty feet from tip to tip. A longer upper wing was his idea on how to make the craft stable. Bray designed the steering and elevating planes. In total the aeroplane weighed 350 pounds and was powered by a 30 horsepower engine. The propeller was seven feet in diameter. This time Bray used spruce instead of bamboo for the frame. Spruce was readily available and offered more strength.

An article in *Aeronautics* described the detailed intricacies and modifications that Bray used. Bray built a very interesting machine of a modified Curtiss type with a great many original details of construction.

Instead of using solid sections for the main beam and outrigger spars, Bray built up spars from two pieces of ½ by 1½ inch spruce held apart by blocks every twelve inches. Wing tip to wing tip it measured thirty feet. The end panels of the lower wing were purposely removed to shorten the lower wing to a twenty-foot spread. The depth of the wings was 5¼ feet, with a curvature of 4½ inches.

The ribs were doubled, running under and over the beams, each strip being ½ by ¾ inches. The strips were held apart by blocks, glued and riveted every twelve inches. The main wings were made in three sections and joined with a steel sleeve, which made it quick and easy to assemble. The uprights were also of novel construction, being made of two pieces of 1 by 1½ inch spruce, separated in the middle by a distance piece of 2½ inches long. The ends were glued and riveted together and fitted into ordinary sockets.

The rear wheels were set about four feet in front of the rear beam, instead of directly under. Bray bought the tires from the Hartford Rubber Company.

The two elevating wings measured 7½ feet with a curvature of one inch and pivoted in front of the main wings. The rear surface was fixed and situated fifteen feet behind the rear spar. The vertical rudder was 30 by 36 inches of ordinary Curtiss construction and operation. The ailerons were ten feet by three inches, covered in canvas, strengthened by double ribs and operated by the usual Curtiss shoulder brace. The outer end of the aileron was supported by a triangle brace from the top wing and a slanting rod from the lower wing. The struts and outriggers were of a peculiar design.

Finally, Bray was ready to test out his flying machine No. 2 on May 9, 1911. This experimental aircraft may have been the lightest craft to fly, and fly it did. Bray brought his craft to Monmouth Park Race Track. He cleared the ground by about ten feet and moved through the air for about 150 feet. Bray maintained an even keel and demonstrated his ability to maneuver the plane as a pilot. The only mishap occurred at landing. One of the landing wheels broke as he touched down.

Willis Wood, Bray's assistant and mechanic, was so encouraged with Bray's success, he decided to build a biplane of his own. Wood modeled it after Bray's machine.

Stephen Willis Wood (1891-1971)

Stephen Willis Wood, who went by Willis, was born in Long Branch, on January 1, 1891, the son of Frank S. Wood and Cornelia. His father was a mason from New York. In 1910, Willis still lived with his parents at 462 West End Avenue. He was an avid sportsman, participating in ice skating contests and getting fined for careless bicycle riding. By 1917, Wood was married and working as a mechanic when he registered for the draft on June 5, 1917. He was stationed at Lakehurst Proving Grounds during WWI.

By 1930, Wood was divorced and worked in a boat yard as a mechanic, living with father and brother in their Long Branch home. In 1940, he lived in Oceanport and worked as a self-employed boat builder. Willis Wood, WWI Veteran, died on April 14, 1971, and is buried as Stephen Willis Wood in Glenwood Cemetery, West Long Branch.

Wood as a Builder of Aeroplanes

Willis Wood built several aeroplanes, the first in 1909. That October, Willis Wood built a twenty-four foot long, four foot wide, four foot tall aeroplane. Just eighteen years old, Willis was considered "somewhat of a genius."

By July 1910, Wood was building another aeroplane and expected to have it done by the fall. It was a Curtiss type plane, old pusher style, where the pilot sits in front of the motor and steers with his shoulders, not his feet.

His first successful flight occurred in 1913 at the Ellwood Park Race Track in Long Branch. Afterwards he placed his plane on exhibition in the Long Branch Casino.

In early February 1914, Wood was testing out his aeroplane at Sea Girt. The plane fell suddenly from about twenty-five feet and crashed on a trial flight. Wood survived the crash with slight injuries but his plane was reduced to kindling.

Wood was injured again in January 1917. He fell against the propeller of his flying machine while the engine was running.

Edward Meyer

Edward Meyer was living in Red Bank during the summer of 1909. In November 1910, Meyer ran for New Jersey Assembly as a Socialist. He was disappointed in receiving only thirty-three votes in a Democratic landslide. It was his first foray into politics. He ran again in the general election held November 5, 1912, as the Socialist candidate for member of the General Assembly State of New York, in Manhattan's Tenth District.

Edward Meyer of Red Bank studied aeroplane design for several years. He began building an aeroplane of his own design during the summer of 1909. Meyer lived on Center Street and worked as a piano delivery man for Frank C. Storck.

The *Wombok*

Storck let Meyer use his piano repair room on Broad Street to assemble the plane in sections. Wing tip to wing tip, it measured thirty feet. The rudder was fifteen feet in back. Meyer's goal was to make it as light as possible. He used bamboo and Oregon spruce to build the framework, which alone weighed just seventy-five pounds. He used piano wire to strengthen the cross sections. With a steering apparatus attached, a Curtiss 25 horsepower engine that he planned to add, and three small bicycle wheels for landing, the total weight was less than 200 pounds. Meyer said his machine was light enough to carry two people. The wings were airproof, rubberized, and "as thin as silk." Meyer called his flying machine the *Wombok*.

Meyer planned to test his biplane on the Red Bank fair grounds. Without the engine yet installed, an automobile would tow him until the plane left the ground. Once airborne, Meyer would cut loose and fly. He told a reporter he would either make a successful flight in his machine or break his neck.

That August, Meyer ran an advertisement soliciting financial backing. He asked for $1000 from any "moneyed man at Red Bank" and he wanted it within twenty-four hours. Meyer tempted readers with the idea that the early money would get the reward. His plan was to patent his design "and have the papers safely in the patent office" before he revealed his machine. He claimed his idea was original, and "by far the finest idea ever brought out." Meyer predicted a new, huge market for flying machines. His patents would become "enormously valuable." He saw possibilities in a "financial way that were far beyond computation." Meyer confidently stated: "My machine is about ready to fly, and it will fly! I have put all my ready cash in the venture and I stake my life on it."

Meyer was especially determined to patent his unique automatic controller device. He may have applied for a patent that August 1909 or later, but a search of US Patents did not reveal any patent granted to him.

A month later, on September 21, 1909, Meyer flew in his machine. He used the Old Monmouth Park Race Track as the testing ground. On a trial a week earlier, Meyer got airborne but the towline broke, causing the plane to crash from about forty feet. Meyer was uninjured and it took less than a week to make repairs and try again

Dr. Edwin Field. *Courtesy of Red Bank Library.*

Despite an unfavorable brisk wind from the east, Meyer proceeded with the test. He arranged with Dr. Edwin Field to be at the track with his car at 2:30 P.M. that afternoon. Dr. Field agreed to tow the plane, but asked Meyer to delay the test until the wind subsided. Meyer agreed to wait until 4:30 P.M.; Dr. Field left to visit a patient.

At 4:30 P.M., the wind was brisker. Dr. Field refused to be a part, but Meyer convinced him to go through with it. "If I should fall and get hurt, the best doctor in New Jersey is there on hand to patch me up." Meyer was ready and willing to take the risk.

In front of about 300 people, they attached the plane to Dr. Field's car. Dr. Field started to drive with Meyer seated in his aeroplane in tow. It rose about four feet in the air then touched down. A second attempt produced the same 'hopping' results, but it was enough for Meyer to go for a full trial flight. They lengthened the towline to 150 feet. With machine in tow, Dr. Field increased to twenty mph and drove around the oval track. Meyer tilted the nose upward and the plane lifted quickly, to a height of about fifty feet. Meyer dropped to about thirty feet to get below the treetops. After a distance of about a half-mile, Meyer descended to land. As he touched down, a gust of wind pushed the plane into a skid. A few spokes broke on one of the landing wheels but other than that there was no harm done. Meyer received applause and an ovation. He was overwhelmed with much handshaking, words of praise and congratulations.

Meyer was especially pleased with how his automatic controller device worked. He was confident that when the engine was added, the device would successfully keep the machine on an even keel. His flight was successful; Meyer had control of his machine in flight.

Trouble with the Law

On October 14, 1909, New Jersey Deputy Marshall Louis G. Beekman arrested Edward Meyer on a warrant for trying to frighten members of a Grand Jury in Kansas, into acquitting a friend. Deputy Beekman took Meyer to Police Headquarters in Newark and locked him up in the jail. His arrest made headlines, partially due to his celebrity status of a successful aviator.

Meyer was held on federal charges of using US Mail to send threatening postcards to jurors empaneled to hear a case against his friend. On May 28, Meyer had written ten identical postcards and had mailed them to ten jurors. He wrote that he was putting their names on the wall of his room, so that he would always be reminded every day. Meyer called them "a fine bunch!" in all capitals and double underlined. In 1909, Meyer's remarks were considered "terms and language of a scurrilous and defamatory character." The trial of United States of America vs. Edward Meyer, was scheduled for November 16. Meyer pled not guilty.

On October 16, Henry N. Supp and Frank C. Storck posted a $1,000 bond. Supp had a clothing store in Red Bank. The case was scheduled to be heard in the District Court of the United States for the District of New Jersey, which met in Trenton. The day before the trial was to convene, Meyer changed his plea to non-vult. The court ordered him to pay a $100 fine and released him.

Letter to the Editor

On April 25, 1910, Meyer wrote a letter to the Editor of the *Red Bank Register*. Meyer wrote that he was unsure when or if he would ever fly again. "Everything I had in the world went into the flying game last summer for patents, experiments, construction and repairs." He claimed to have experienced three falls, each time smashing the aeroplane. Meyer said his monoplane with warping tips was rearranged into an entirely new racing machine. He said he had the patents ready for the *Wombok Second*. If he could finance the patents, it would give Red Bank a red star on the map. He said the aeronautical magazines in England, Germany and France would publish his complete story of his experiments and results that month.

Steamer *Sea Bird. Courtesy of Red Bank Library.*

In July 1910, Meyer shipped his biplane by the steamboat *Sea Bird* to Coney Island. George C. Tilyou had hired Meyer as an attraction and agreed to finance the project. Meyer planned to finish reassembling his plane in the Pavilion of Fun at Steeplechase Park, then make exhibition flights in his biplane. Tilyou hired Meyer to give exhibition flights for eight weeks while he was there. Meyer was a former balloonist and took lessons in Atlantic City from pilots Glenn H. Curtiss and Walter Brookins. Meyer's plane was similar in design to Wright's biplane, measuring thirty feet from tip to tip, but with several changes by Meyer.

Morris F. Morris (1870-1959)

Morris Frank Morris was born on July 14, 1870, in Philadelphia, Pennsylvania. His second marriage was to Annie Fitzgerald on April 14, 1900, in Toms River, New Jersey. In 1910, Frank and Annie lived at 68 Ocean Avenue in Long Branch, with their three children, Helen, Frank and William. Morris worked as a house painter and steeplejack. In 1912, they lived at 3 Madison Avenue. Frank & Annie celebrated their golden wedding anniversary April 14, 1950. Frank Morris died March 6, 1959, in Elberon, New Jersey.

Morris Frank Morris, *left,* and family. *Courtesy of Jolie Morris.*

Morris Builds a Flying Machine

On October 2, 1912, it became public news that Frank Morris was building a flying machine. Morris had been working on his aeroplane for several months. His machine incorporated some features of his own design and he was ready for a trial flight.

On September 24, 1913, "Steeplejack" Morris took a flight in his aeroplane. Harold E. Powers agreed to let Morris use his farm in Shrewsbury for the test. With the machine positioned on the top of a hill, Morris started the engine. The machine rose off the ground.

When it reached a height of about eight feet, one of the wings broke off and the machine fell to the ground. Despite the fall, Morris was encouraged and was eager to make repairs and plan another attempt soon.

A month later, on October 24, 1913, Morris was testing his monoplane on Ocean Avenue in Long Branch but ran into an automobile. The collision damaged the propeller and broke some of the wire supports.

Morris tried again to fly on Thanksgiving Day, November 25, 1915. The Hollywood Golf Course in Deal allowed him to use their grounds. Unfortunately, before leaving the ground the plane got loose from Morris and his assistant. One of the wings was broken off, leaving his airship a wreck.

According to family lore, Frank suffered a leg injury which caused him a lot of pain. His descendants believed that Frank built his aeroplane with the hopes of selling it to the military. Frank thought the US Army would find his designs of good use. He contacted the Army and they agreed to see a demonstration. Frank hired a test pilot, but when the test pilot showed up drunk, Frank flew the plane himself. Unfortunately, Frank was inexperienced and crashed his machine.

Kimmerland Builds an Aeroplane

Peter Kimmerland, aka Kimmerling, was a German immigrant who settled in Long Branch and worked as an upholsterer. In August 1909, the *Red Bank Register* reported that he invented a flying machine. Kimmerling claimed it was not like any other. Two months later, the local newspapers reported that Kimmerland's trial flights failed.

Morford Crashes

Leon A. Morford was an aviator of Allentown. While flying one day in the spring of 1915, he crashed from about 200 feet. Morford suffered serious injuries. His recovery was slow, but after a few months he was able to get up and walk

-----

Howard S. Borden (1877-1950)

General Howard S. Borden was a millionaire who enjoyed many hobbies including flying. In July 1916, Borden bought a Burgess-Dunne hydroaeroplane. The eight-cylinder engine could produce 100 horsepower. It had three pontoons and seats for an aviator and a passenger. Borden built a hangar on his summer estate on the Navesink River to shelter his aeroplane. Borden commuted to work from his summer home in Rumson to New York City via his yacht, the *Sovereign*, but planned to use his hydroplane. Borden had his chauffer Orrin G. Soule trained in operating the machine so Soule could teach him how to operate it. Soule lived at the Borden estate in Oceanic.

Borden's flying boat, by Burgess-Dunne, on the Shrewsbury River, circa 1916. *Courtesy of Nancy Bryne Phillips, and Roberta Van Anda.*

Patrick J. Byrne (1896-1979)

As a youngster, Patrick Byrne learned to fly using Borden's aeroplane. During WWI, Byrne received Enlisted Pilot License No. 11. A life-time pilot, Byrne spent forty-two years in the US Navy. Byrne participated as a stunt pilot in the late 1929 air show at Red Bank Airport.

Patrick J. Byrne. *Courtesy of Nancy Bryne Phillips, and Roberta Van Anda.*

Schroeder's Aeroplane (1888-1979)

Edward J. Schroeder of Manasquan and his brother Charles J. Schroeder were building an aeroplane in July 1910. Schroeder had set motorboat speed records in his *Dixie II* but decided to try his hand at flight. Their design would allow the craft to take off and land on water. The Schroeder brothers planned to test it on the Manasquan River after they completed its construction.

Sources

1. Charles J. Hendrickson

Aeronautics, "The Stability of Aeroplanes," by C. J. Hendrickson, Vol. III, No. 2, p.22, August 1908.
Aeronautics, "All Ready for Aeronautic Society Exhibition," Vol. V, No. 1, p.17, July 1909.
Bartlett, Kathleen, descendant.
Bell Laboratories Record, Vol. XXIV, No. 11, p.30, February, 1946.
Brooklyn Eagle, "Aero Club of America Names Special Contest Committee for June 5 Event," p.18, May 23, 1909.
Matawan Journal, "Farmer's Son Makes Flying Machine," p.6, April 23, 1908.
New York Times, "Aeroplanes for the Army," p.5, October 2, 1908.
New York Times, "Girl Dies in Stunt Boarding Airplane from Running Auto," p.1, October 5, 1921.
Ocean Grove Times, "Within Monmouth County," p.4, November 26, 1910.
Red Bank Register, "Made a Flying Machine," p. 1, April 22, 1908.
Red Bank Register, "Middletown Village News," p.16, March 17, 1909.
Red Bank Register, "Charles Hendrickson Flies Fifteen Yards in his Airship," p.15, March, 31, 1909.
Red Bank Register, "Actress Killed in Fall," p.15, October 5, 1921.

2. America's Greatest Aviation Meet

Asbury Park Press, "All is Ready for First Flight of Famous Birdmen Tomorrow," p.1, August 9, 1910.
Asbury Park Press, "Crowds at Aviation Field Despite Overcast Skies," p.1, August 10, 1910.
Asbury Park Press, "Wreck of Biplane; Hurt 12 in its Fall," p.1, August 11, 1910.
Asbury Park Press, "Brookins Always Brave says Wright," p.1, August 11, 1910.
Asbury Park Press, "Aerial Program This Afternoon," p.4, August 11, 1910.
Asbury Park Press, "Governor's Day Draws Great Crowds to the Aviation Field," p.1, August 12, 1910.
Asbury Park Press, "Drexel Wrests Altitude Record from Brookins," p.1, August 12, 1910.
Asbury Park Press, "Aeronaut Falls 4000 Feet to Awful Death," p.1, August 12, 1910.
Asbury Park Press, "Brookins Will Try for Drexel's Record," p.9, August 13, 1910.
Asbury Park Press, "Birdmen to Fly Despite Weather," p.1, August 14, 1910.
Asbury Park Press, "Peppin Would Fly," p.1, August 14, 1910.
Asbury Park Press, "Wright Talks of Flight of Drexel," p.1, August 14, 1910.
Asbury Park Press, "Local Man Inventor of a New Air Craft," p.1, August 14, 1910.
Asbury Park Press, "Brookins Eager to Fly, Will Make Flight Tomorrow," p.1, August 16, 1910.
Asbury Park Press, "Half Gale Prevents Air Flights Today," p.1, August 17, 1910.
Asbury Park Press, "Makes Flight in Half Gale," p.10, August 18, 1910.
Asbury Park Press, "Cross Country Flight to Sea Girt Scheduled Today," p.1, August 20, 1910.
Asbury Park Press, "Johnstone and Hoxsey Make Thrilling Night Flights," p.5, August 20, 1910.
Asbury Park Press, "Aviation Meet Receipts $25,000," p.2, August 21, 1910.

Asbury Park Press, "Motor Balks Johnstone's Attempt at Height Record," p.1, August 21, 1910.
Asbury Park Press, "Boy Hit by Biplane Asks $15,000 Damages," p.5, August 22, 1910.
Asbury Park Press, "Aviators Undaunted by Wind that Rocks Planes," p.6, August 24, 1910.
Asbury Park Press, "Breeze Hinders the Aviators," p.4, August 25, 1910.
Asbury Park Press, "Hoxsey Makes Flight Over Sea and Along Asbury's Beach-front," p.1, August 27, 1910.
Asbury Park Press, "Thrilling Flights Bring Aviation Meet to Close," p.1, August 28, 1910.
Monmouth County Archives, Incorporations, Book G, p.52, Aero and Motor Club of Asbury Park, July 20, 1910.
New York Times, "Aviators to Hold a Meet for Ten Days in Mid-August," p.5, July 24, 2910.
New York Times, "75,000 Visitors Expected at Opening Day," p.14, August 6, 1910.
New York Times, "Aviation Meet at Asbury Park," p.2, August 6, 1910.
New York Times, "Aviation Meet at Asbury Park," p.2, August 7, 1910.
New York Times, "Asbury Park Meet Today," p.20, August 10, 1910.
New York Times, "Brookins Falls, Eight Injured," p.1, August 11, 1910.
New York Times, "Fall of Brookins Due to his Daring," p.3, August 12, 1910.
New York Times, "6,000 Feet to Death for a Parachute," p.1, August 13, 1910.
New York Times, "Johnstone Flies 5,000 Feet in Air," p.1, August 13, 1910.
New York Times, "Wright Biplane Will Cary Five," p.4, August 16, 1910.
New York Times, "Wright Pupils Fly in Heavy Storm," p.3, August 17, 1910.
New York Times, "Aeroplane Crashes into Automobiles," p.2, August 18, 1910.
New York Times, "Night Ascension for Aeroplanes," p.2, August 18, 1910.
New York Times, "Airmen Play Tag with Moonbeams," p.16, August 20, 1910.
New York Times, "The New Wright Five-passenger Biplane," p.12, August 21, 1910.
New York Times, "Goes Visiting in Airship," p.2, August 21, 1910.
New York Times, "Aviator Brookins in Surprising Feats," p.6, August 22, 1910.
New York Times, "Brookins Sets Mark for a Quick Turn," p.6, August 24, 1910.
New York Times, "Hoxsey Makes Ocean Trip," p.8, August 26, 1910.
Ocean Grove Times, "Aviation Meet is On," p.1, August 13, 1910.
Ocean Grove Times, "Aerial Flights Continue," p.1, August 20, 1910.
Ocean Grove Times, "Hurt by Airship, Sues," p.1, August 27, 1910.
Red Bank Register, "Aviation Meet Assured," p.3, July 27, 1910.
Red Bank Register, "Motor for Airplane," p.5, August 3, 1910.
Red Bank Register, "Coming Aviation Meet," p.9, August 3, 1910.
Red Bank Register, "Aviation Meet," p.2, August 10, 1910.
Red Bank Register, "Asbury Park Man on Governor's Staff," p.5, August 17, 1910.
Red Bank Register, "Aeroplane Strikes Ground," p.7, August 17, 1910.
Red Bank Register, "Balloonist Drops to Death," p.13, August 17, 1910.
Red Bank Register, "Takes Observations," p.1, August 24, 1910.
Red Bank Register, "Airship Goes Through Fence," p.3, August 24, 1910.
Red Bank Register, "Aviation Meet Continued," p.9, August 27, 1910.
Red Bank Register, "Damage Suit for Aeroplane Injury," p.3, August 31, 1910.
Red Bank Register, "Aviators Win Suit," p.5, July 3, 1912.
Red Bank Register, "Aeroplane Hits a Tree," p.10, August 27, 1913.
US Patent Office, Patent No. 970616, September 20, 1910.

3. Monmouth Aviation Patents
Walling
Asbury Park Press, "Patents Airship That Won't Upset," p.1, September 29, 1910.
Asbury Park Press, "Walling Tries His Aeroplane Motor," p.9, October 20, 1910.
Asbury Park Press, "Aeroplane with Movable Wings May Upset Science of Flying," p.1, November 2, 1910.
Asbury Park Press, "Walling's Aeroplane Refuses to Go Aloft," p.3, November 23, 1910.
The Globe and Commercial Advertiser, "Tinton Falls Will Soar to More Fame; Walling Airship and the Brothers Who Built It," November 8, 1910.
New York Herald, "Sold His Horse to Build an Airship."
O'Brien, Sally, descendant.
Ocala (Florida) Evening Star, "Thomas M. Walling," p.3, December 23, 1935.
Ocean Grove Times, "Second Business Show in the Casino Next Week," p.4, January 20, 1922.
Red Bank Register, "Cook-Walling," p.5, December 6, 1899.
Red Bank Register, "A New Flying Machine," p.1, September 28, 1910.
Red Bank Register, "Flying Trip Delayed," p.1, October 5, 1910.
Red Bank Register, "Crowd Watches Motor at Work," p.1, October 19, 1910.
Red Bank Register, "Ready for Trial Flight," p.9, November 2, 1910.
Red Bank Register, "Airship Given a Tryout," p.12, November 19, 1910.
Red Bank Register, "Flying Machine on Show," p.9, August 23, 1911.
Red Bank Register, "Thomas M. Walling Sells Farm to William Van Pelt," p.18, September 7, 1921.
Red Bank Register, "Thomas M. Walling," p.13, December 26, 1935.
Red Bank Register, "Thomas M. Walling," p.11, January 2, 1936.
US Patent Office, Patent No. 1,004,944, October 3, 1911.
Wilshire's Magazine, "A Farmer's Aeroplane," Vol. 14-17, p.33 & 85, September, 1910.
Winters, Dick, descendant.

Moore
Aeronautics, Vol. X, p.181, May-June 1912.
Davis, Kris, descendant.
Nelson, Debbie, descendant.
Seattle Times, obituary, p.49, April 8, 1963.
US Patent Office, Patent No. 1,019,987, March 12, 1912.

Hurst
Aerial Age Weekly, "Evansville is Center of Aviation Activity," Vol. XI, No. 7, p.217, April 26, 1920.
Aerial Age Weekly, "Hurst Company to Build Plant," Vol. XI, No. 20, p.676, July 26, 1920.
Monmouth County Archives, Corporations, Hurst Aircraft Corp., Book I-168, February 14, 1917.
New York Times, "Navy Wants Seaplanes," p.11, February 28, 1915.
New York Times, "Fall to Death in Liberty Plane," p.11, January 17, 1918.
Red Bank Register, "He Kept the Premiums," p.9, July 15, 1914.
Red Bank Register, "New Aeroplane Company," p.3, February 7, 1917.
Red Bank Register, "Airship Company Incorporated," p.9, March 7, 1917.
Trenton Evening Times, "Moveable Wings on Hydro-Aeroplane," p.10, November 5, 1914.

US Patent Office, Patent No. 994,104, May 30, 1911.
US Patent Office, Patent No. 1,424,143, July 25, 1922.

Stevens
Red Bank Register, "Rumson Man Is Inventor," p.23, April 18, 1928.
Red Bank Register, "Patent on Parachute," p.1, February 27, 1929.
US Patent Office, Patent No. 1,702,422, February 19, 1929.

Fox
US Patent Office, Patent No. 1,037,977, September 23, 1923.

Havens
US Patent Office, Patent No. 1,165,168, December 21, 1915.

4. Gala at Rest Hill
Aero Club of America Bulletin, "The Alluring Sport of Flying," by Robert J. Collier, Vol. I, No. 5, p.3, June 1912.
Aeronautics, Vol. VIII, No. 1, p., 129, 1911.
Aeronautics, "New Burgess Flying Boat," Vol. XIII, No. 2, p.48, August 1913.
Aero and Hydro, "An Air-Living Sportsman," Vol. VII, No. 13, p.69, December 27, 1913.
Aero and Hydro, "Collier's Burgess Makes Fast Time," Vol. VII, No.14, p.166, January 3, 1914.
Aero and Hydro, Vol. VII, No.24, "Aero-Skimmer Built for Collier," p.303. March 14, 1914.
Aeronautics, Vol. XV, No.1, "Flying Sloane Boat," p.8, July 15, 1914.
Aero, America's Aviation Weekly, "Collier Holds Private Meet," Vol. III, No. 5, p.98, November 4, 1911.
Aircraft, "The Burgess Flying Boat Built for Robert J. Collier," by F. H. Russell, Vol. IV, No. 7, p.164, September 1913.
Asbury Park Press, "Collier Flies His Biplane to Lakewood," p.1, October 17, 1911.
Matawan Journal, "Mr. Collier Had Narrow Escape," p.1, October 19, 1911.
New York Times, "Tom Sopwith Arrives," p.C9, May 7, 1911.
New York Times, "Follow Hunt in Air at Collier's Fete," p.15, October, 15, 1911.
New York Times, "Follow Hunt In Air at Collier's Fete," p.15, October 15, 1911.
New York Times, "Aero Club Would Try 7 Different Routes to Europe," p.5, June 24, 1918.
New York Times, "R.J. Collier Dies at Dinner Table," p.13, November 9, 1918.
Red Bank Register, "Gathering of Farmers," p.1, October 11, 1911.
Red Bank Register, "Aviation Meet," p.31, December 13, 1911.
Red Bank Register, "Great Day at Rest Hill," p.9, June 5, 1912.
Red Bank Register, "Aeroplane Falls in Bay," p.1, August 28, 1912.
Red Bank Register, "Robert J. Collier Dead," p.9, November 13, 1918.

5. Aeromarine Plane and Motor Company
Aerial Age, Vol. 1, No. 4, "Aeroplanes Invaluable in European War," p.87, April 12, 1915.
Aerial Age, Vol. II, No. 19, p.459, January 24, 1916.
Brooklyn Eagle, "Janney, Canada's Flying Ace," p.27, June 5, 1927.
Matawan Journal, "Aeroplane Plant at Keyport," p.8, May 16, 1916.

Matawan Journal, "Buys the Output of Aeromarine Company," p.1, March 18, 1920.
New York Times, "Janney Now Plans Hop for London," p.6, June 30, 1927.
Red Bank Register, "News from Keyport," p.14, April 5, 1916.
Red Bank Register, "News from Keyport," p.5, April 26, 1916
Red Bank Register, "New Aeroplane Factory," p.14, May 24, 1916.
Red Bank Register, "Keyport's Industries," p.2, June 7, 1916.
Red Bank Register, "Safe and Sane Fourth," p.12, June 21, 1916.
Red Bank Register, "Aeroplane Damaged," p.4, July 12, 1916.
Red Bank Register, "News from Keyport," p.15, August 2, 1916.
Red Bank Register, "Lincroft News," p.8, June 26, 1918.
Red Bank Register, "Help Wanted," p.14, November 26, 1919.

Boland
Aero & Hydro, "Boland," Vol. V, No. 18, p.331, February 1, 1913.
www.earlyaviators.com/ebolandf.htm

Janney
Aerial Age Weekly, "Aeroplanes Invaluable in European War," Vol. 1, No. 4, p.87, April 12, 1915.
Brooklyn Eagle, "Janney, Canada's Flying Ace, May Lead a Dozen Planes in Race to London on June 30," p.27, June 5, 1927.
Flying, advertisement, Volume VI, No. 4, p.310 May, 1917.
Gray, Larry, *We are the Dead,* p.198, 2001.
Monmouth County Archives, Corporation and Trade Name Certificates, Janney Aircraft Co., Box 813, May, 29, 1916.
New York Times, "Janney Now Plans Hop for London on July 10," p.6, June 30, 1927.

Uppercu
Aerial Age Weekly, Vol. II, No. 19, p.459, January 24, 1916.
Aerial Age Weekly, "Uppercu, President of ACC One of America's Most Enthusiastic Supporters of Aviation," Vol. XV, No. 20, p.545, November 1922.
Aviation, "Aero Chamber of Commerce Elects Officers," Vol. XIII, No. 11, p.313, August 1922.
Air Service Journal, "Ingliss M. Uppercu," Vol. I, No. 11, p.340, September 17, 1917.
Daily Home News (New Brunswick), "To Shoot from Air at Sea Girt, as they do in Europe," p.3, August 24, 1916.
Matawan Journal, "Aeroplane Plant for Keyport," p.8, March 16, 1916.
Red Bank Register, "Aeromarine Company to buy Brickyard Property," p.18, March 22, 1916.
Red Bank Register, "Aeroplane Factory Brings Prosperity to the Town," p.1, September 26, 1917.
Red Bank Register, "News from Keyport," p.5, October 10, 1917.
"The Log of an Aeromarine, a Modern Adventure in Pathfinding," Aeromarine Plane & Motor Company, Keyport, New Jersey, 1920.
US Tax Court, Chilton vs. Commissioner, Docket No. 91332, June 21, 1963.

Robinson
New York Times, "Hydro-Aeroplanes at Monte Carlo," p.5, March 25, 1912.

DeGiers

Aero and Hydro, "Among the Aviators," p.173, May 18, 1912.
Chevalier, George M. "This Was Panama," BookSurge Publishing, 2002.
US Patent Office, Patent No. 1,384,308, July 12 1921.

6. Casey Brothers

Asbury Park Press, "Woman Is Killed, Brother Maimed by Pilotless Airplane," p.1, January 16, 1922.
Brooklyn Eagle, "Airplane Kills One, Injures 50 Others," p.7, January 16, 1922
Brooklyn Daily Eagle, "5 Young Explorers to Sail Dec. 5 for Andes Air Survey," p.19, November 24, 1930.
Monmouth County Archives, Incorporations, Book N, Airview, Inc., p.76, June 30, 1926.
Monmouth County Archives, Incorporations, Book O, Red Bank Aero Club, p.129, October 11, 1927.
Monmouth County Archives, Incorporations, Book P, Airview Flying Service, Inc., p.2, July 11, 1928.
Monmouth County Archives, Incorporations, Red Bank Airport, Book P, p.296, February 13, 1929.
Monmouth County Archives, Incorporations, Red Bank Air Races, Inc., Book P, p.420, March 26, 1929.
New York Times, "Wild Airplane Kills Woman, Maims Man," p.1, January 16, 1922.
Red Bank Register, "A Millionaire Hobby," p.12, July 12, 1916.
Red Bank Register, "Enlists in Signal Corp," p.13, February 27, 1918.
Red Bank Register, "Will Learn to Fly in Texas," p.11, May 15, 1918.
Red Bank Register, "James Casey 'Over There,'" p.14, July 24, 1918.
Red Bank Register, "Brothers Get Promotions," p.9, October 9, 1918.
Red Bank Register, "Bought an Aeroplane," p.1, July 30, 1919.
Red Bank Register, "Trips in an Aeroplane," p.9, August 13, 1919.
Red Bank Register, "Cabbage by Air Route," p.1, September 24, 1919.
Red Bank Register, "Flew to Atlantic City," p.9, October 8, 1919.
Red Bank Register, "A New Aviator," p.9, October 22, 1919.
Red Bank Register, "Left the Earth," p.4, October 29, 1919.
Red Bank Register, "Will Sell Air Machines," p.1, November 5, 1919.
Red Bank Register, "Aerial Advertising," p.9, November 12, 1919.
Red Bank Register, "Fireworks from an Airship," p.1, January 7, 1920.
Red Bank Register, "Big Railroad Strike is on," p.1, April 14, 1920.
Red Bank Register, "Up through the Clouds," p.1, May 5, 1920.
Red Bank Register, "Bought a Flying Boat," p.11, June 2, 1920.
Red Bank Register, "Using Airplanes for Advertising," p.12, June 16, 1920.
Red Bank Register, "Up in the Air with Casey," p.11, July 7, 1920.
Red Bank Register, "Back from Aerial Trip," p.1, October 12, 1921.
Red Bank Register, "Flew over New York," p.1, November 2, 1921.
Red Bank Register, "Killed by an Aeroplane," p.11, January 18, 1922.
Red Bank Register, "Made $500 for Injured Man," p.1, May 17, 1922.
Red Bank Register, "Too Many Automobiles," p.13, October 29, 1924.
Red Bank Register, "In Flames in the Air," p.9, November 25, 1924.
Red Bank Register, "View from the Clouds," p.14, May 5, 1926.
Red Bank Register, "Flying Field Opened," p.13, May 4, 1927.
Red Bank Register, "Red Bank's," p.1, May 25, 1927.
Red Bank Register, "Two Planes on Field," p.14, October, 2, 1927.
Red Bank Register, "An Airplane Club," p.25, October 19, 1927.

Red Bank Register, "Red Bank has Monmouth's First Aerial Club," p.16, October 26, 1927.
Red Bank Register, "Aviation Field for Red Bank Now Seems a Certainty," p.4, March 7, 1928.
Red Bank Register, "Busy Days for Aviator," p.1, April 4, 1928.
Red Bank Register, "Jack Casey in the Movies," p.1, April 24, 1929.
Red Bank Register, "Building an Airplane," p.1, September 12, 1928.
Red Bank Register, "Red Bank's New Light," p.1, September 12, 1928.
Red Bank Register, "Casey Helps in Emergency," p.26, May 22, 1929.
Red Bank Register, "Red Bank Air Races," p.1, June 26, 1929.
Red Bank Register, "More Room for Airport," p.5, September 4, 1929.
Red Bank Register, "J. F. Casey, Aviation Pioneer," p.2, May 29, 1964.
Red Bank Register, "John F. Casey," p.2, June 1, 1964.

7. De Luxe Air Services
Morgan
Aerial Age, "The Aircraft Trade Review," Vol. XII, No. 12, p.342, December 6, 1920.
Aerial Age, "How Safe is Flying?" Vol. XV, No. 4, p.77, April 3, 1922.
Aircraft Journal, "Stuart Morgan Loading Merchandise in an Airplane," Vol. 4, p.9, June 28, 1919.
Flying, "Aerial Service at Asbury Park, NJ," Vol. IX, No. 6, p.391, July, 1920.
Goodrich, Peggy, *History of Shark River Hills*, p.9, 1984.
Monmouth County Archives, Incorporation Records, De Luxe Air *Ocean Grove Times*, "Asbury Park News Notes," p.2, April 1, 1921.
Ocean Grove Times, "Asbury Park News Notes," p.8, April 22, 1921.
Ocean Grove Times, "Business Show was opened with a Rush," p.1, January 27, 1922.
Service, Inc., Book J, p.111, March 29, 1929.
Red Bank Register, "An Aviation Club," p.1, April 16, 1919.
Red Bank Register, "Airplane Smashed," p.2, June 18, 1919.
Red Bank Register, "Airplane Sold," p.5, July 9, 1919.
Red Bank Register, "Has New Flying Machine," p.7, July 23, 1919.
Red Bank Register, "A Flying Job," p.12, April 27, 1921.
Red Bank Register, "A Wedding in the Clouds," p.7, June 22, 1921.
Red Bank Register, "Asbury Park's Business Men's Show," p.8, January 18, 1922.
Tobin, Ardy, descendant.
www.earlyaviators.com/emorgan.htm

8. Monmouth Builders and Flyers
Bray
Aeronautics, Vol. V, No. 216, p.177, November 1910.
Aeronautics, "Flies with 2-Cylinder Motor," Vol. VIII, No. 1, p.215, July 1911.
History of Monmouth County, New Jersey, 1664-1920, Vol. 3, p.322.
New York Times, "Business Opportunities," p.47, October 8, 1922.
Ocean Grove Times, "In Monmouth County," p.8, November 13, 1909.
Ocean Grove Times, "Long Branch Aviator," p.8, December 18, 1909.
Ocean Grove Times, "Long Branch Aviator," p.1, September 24, 1910.
Red Bank Register, "Northam-Bray marriage," p.5, September 22, 1896.
Red Bank Register, "In and Out of Town," p.16, September 1, 1897.
Red Bank Register, "Long Brancher Making Aeroplane," p.3, November 24, 1909.
Red Bank Register, "Airship Stuck in the Mud," p.14, June 22, 1910.

Red Bank Register, "Airship Sticks in the Mud," p.7, June 23, 1910.
Red Bank Register, "Building New Bi-Plane," p.5, July 20, 1910.
Red Bank Register, "New Biplane Completed," p.5, September 28, 1910.
Red Bank Register, "Bray Up in the Air," p.6, May 10, 1911.
US Patent Office, Patent No. 880,714A, March 3, 1908.

Wood
History of Monmouth County 1664-1920, Vol. 2, p.114.
Ocean Grove Times, "In Monmouth County," p.8, November 13, 1909.
Red Bank Register, "Items of News," p.5, July 20, 1910.
Red Bank Register, "Another Aviator," p.5, July 27, 1910.
Red Bank Register, "Big Day on the Ice," p.1, February 19, 1913.
Red Bank Register, "Airplane Goes Down," p.2, February 11, 1914.
Red Bank Register, "Hot Water Heated Bus," p.6, October 6, 1915.
Red Bank Register, "Amateur Aviator Hurt," p.6, January 17, 1917.
Red Bank Register, "Home from Long Auto Trip," p.10, August 31, 1921.
Red Bank Register, "Highlands Boat Races," p.24, July 18, 1928.

Meyer
Brooklyn Eagle, "Gossip Stage," p.17, July 27, 1910.
Brooklyn Eagle, "Amusements at Coney Island," p.15, August 7, 1910.
Matawan Journal, "Red Banker's Aeroplane a Success," p.6, September 23, 1909.
New York Daily Tribune, "Biplane Flies, Towed Like a Kite," p.5, September 22, 1909.
New York Times, "Aviator Put in Jail," p.20, October 15, 1909.
Red Bank Register, "A New Flying Machine," p.1, August 4, 1909.
Red Bank Register, "Kite Flying at the Fair," p.9, August 25, 1909.
Red Bank Register, "Cash Wanted to Promote Aeroplane," p.13, August 25, 1909.
Red Bank Register, "Meyer Makes a Flight," p.9, September 15, 1909.
Red Bank Register, "Made a Half-Mile Flight," p.9, September 22, 1909.
Red Bank Register, "Ed Meyers Arrested," p.3, October 20, 1909.
Red Bank Register, "Fined $100," p.12, November 17, 1909.
Red Bank Register, "Installed New Lighting," December 15, 1909.
Red Bank Register, "Letter to the Editor," p.4, April 27, 1910.
Red Bank Register, "May Fly at Coney Island," p.9, July 20, 1910.
Red Bank Register, "Stalled in an Auto," p.1, October 5, 1910.
Red Bank Register, "A Democratic Landslide," p.9, November 9, 1910.

Morris
Long Branch Daily Record, "Mark Golden Wedding 50th Anniversary," p.30, April 14, 1950.
Morris, Jolie, descendant.
Ocean Grove Times, "Within Monmouth County," p.8, October 26, 1912.
Red Bank Register, "Falls from Smokestack," p.22, November 30, 1910.
Red Bank Register, "Building a Flying Machine," p.2, October 2, 1912.
Red Bank Register, "Aeroplane Takes a Tumble," p.13, September 24, 1913.
Red Bank Register, "Monoplane Hits Auto," p.2, October 29, 1913.
Red Bank Register, "Flag Pole Painted," p.3, May 27, 1914.
Red Bank Register, "Airship Wrecked," p.6, December 1, 1915.

Kimmerland

Long Branch Daily Record, "Three Aviators in Long Branch," p.1, December 10, 1909.
Red Bank Register, "Invents Flying Machine," p.3, August 4, 1909.

Morford
Red Bank Register, "Aviator Recovering," p.7, June 2, 1915.

Schroeder
New York Times, "E. J. Schroeder Gives up Motor Boating," p.5, July 10, 1910.
Red Bank Register, "To Try Aeroplane on River," p.5, July 27, 1910.

Byrne
Philip, Nancy Byne, descendant.

Index

Ackerson, Cornelius, 86
Aero Club, of
  America, 66, 85
  Asbury Park, 13, 14, 15, 23,
    34, 35
  Atlantic City, 113
  Camden, 113
  New Brunswick, 113
  Newark, 113
  Pennsylvania, 33
  Red Bank, 97, 113, 114, 116,
    121
Aeromarine Plane & Motor Co.,
  83, 85, 86, 87, 89, 90, 91, 96,
  98, 102, 108
Aitken, William H., 4
Allen, George K., Jr., 34
Antoinette, 127
Anzani, Allesandro, 77
Appleby, T. Frank, 25, 26
Atkins, Charles A., 16
Atwater, Benjamin L., 113
Aumack, Theodore, 83
Aymar, John W., 16

Bailey, Mary, 81, 100
Barker, William H., 97
Barrymore, Ethel, 91
Beekman, Louis G., 133
Benjamin, Franke, 100
Berry, William A., 16
Bjornstad, Andreas, 58
Blakeley, Harold W., 57
Blume, George L., 62
Boeing, 49, 50
Boland, Frank E., 82, 85
Boland, Joe, 82
Borden, Howard S., 102, 106,
  137
Bradley, James A., x
Bradway, C., 62
Bray, Gertrude, 118
Bray, Harriet, 118
Bray, Joseph, 118

Bray, Nettie, 126
Bray, Ralph L., 118, 126, 127,
  128, 129
Briggs, Frank O., 25, 26
Brindley, Richard, 24
Brookins, Walter A., 15, 16, 19,
  21, 22, 23, 24, 28, 30, 31, 35,
  36, 78, 134
Brooks, John B., 9
Brunner, Adam, 114
Buckingham, Thomas A., 58
Burgess Co., 70, 76, 78, 102,
  137
Burnett, George, 23
Burnett, Harry B., 23
Byrd, Richard E., 112, 113
Byrne, Patrick J., 137, 138

Carberry, Joseph E., 106
Casey, James H., 81, 100, 102,
  104, 106, 109, 110
Casey, James, Sr., 81
Casey, John F., 97, 81, 100,
  101, 102, 103, 104, 105, 106,
  107, 108, 111, 112, 113, 114,
  115, 116, 122
Casey, Madeline, 101
Casey, Mary, 100
Casey, Nelly, 100
Casey, William J., 111
Chanute, Octave, 4, 5, 6
Childs, Harry B., 84
Chilton, Emmie, 97
Chilton, Roland, 97
Clayton, Clarence, 53
Clayton, Harry H., 110
Coffyn, Frank, 16, 22, 26, 28,
  30, 31, 35, 66, 72, 76, 77, 78
Cole, Howard, 108
Collier, Peter F., 66
Collier, Robert J., 66, 67, 68, 69,
  70, 72, 73, 74, 75, 76, 77, 78,
  96, 101
Colman, Andrew R., 48

Conley, Lawrence, 109, 110, 111
Conover, Charles E., 68
Conover, Henry, 68
Conover, J. Clark, 103, 104, 105, 107, 108, 111, 113, 114, 122
Cook, Charles E., 23, 24
Cook, Robert L., 42
Curtiss Aeroplane Co., 9, 34, 96, 97, 101, 104, 106
Curtiss, Glenn, 91, 94, 134

Davis, Madeline, 45
De Luxe Air Service, 120
DeBow, Richard E. A., 58
DeGiers, Clarence A., 96, 97
Demoiselle, 6, 7
Denegar, Harold E., 16
Doremus, Thomas P., 115
Drexel, J. Armstrong, 31, 32, 33
Duffy, John, 101
Dungan, Nelson Y., 23
Dunning, Joseph, 34, 35

Edison, Thomas A., 28
Edwards, Clifton V., 54
Evans, Victor, 61
Evenson, Christine, 49
Fell, Abercrombie, Mrs., 16, 22

Ferguson, Will O., 58
Ferris, Martin L., 62
Field, Edwin, Dr., 23, 72, 110, 132
Fitzgerald, Annie, 135
Fort, John F., Gov., 25, 26
Fox, Ralph P., 61
Fromme, Isaac, 24
Fromme, Warren, 24
Frost, Margaret H., 16

Garretson, Frank, 69, 70
George Washington, 57
Gorsuch, Maurice, 23
Gray, George, 36

Guerin, Claude V., 25, 26

Hackman, Henry D., 31
Hanigman, James K., 113
Harding, William B., 115
Hardy, George W., 61
Hare, James H., 67, 74
Havens, John T., 62
Haycock, Scott R., 62
Hazelhurst, Leighton W., Jr., 76
Hazleton, Herbert, 84
Healy, Arthur, 24
Hearst, Randolph, x
Heinrichs, Louis R., 53
Hendrickson, Charles J., 1, 2, 3, 4, 5, 6, 8, 9, 10
Hendrickson, John S., 1
Henenberg, Herbert, 113
Herbert, Henry L., 67
Hettrick, Clarence E., 26, 119
Hild, Frederick C., 36
Hill, William B., 96
Hoffman, Harold G., 112
Hoffman, Lillian, 112
Hoiriis, Holger, 114, 115
Hounihan, Anna C., 109, 110
Hounihan, Grace, 109
Hounihan, Harold, 109
Hounihan, Raymond, 109
Howell, Benjamin F., 26
Howland, Harry W., 54
Hoxsey, Archibald, 16, 18, 22, 26, 31, 32, 33, 34, 36
Hubbard, James, 70, 72
Hurst, William C., 54, 55, 56, 57, 58

Irwin, Charles P., 108, 110

Janney, Elizabeth, 83
Janney, Ernest L., 83, 84, 85
Janney, William, 83
Johnson, Garret I., 111
Johnson, George R., 111, 113, 115
Johnson, Howard A., 61

Johnson, John M., 102, 103
Johnstone, Ralph, 16, 21, 27,
  28, 31, 32, 36
Jones, William T., 51

Kean, John., 26
Kelley, John H., 113
Kerr, George N., 54
Kimmerland, Peter, 136
Kinmonth, J. Lyle, 16
Kirkbride, Samuel, 26
Knabenshue, A. Roy, 12, 13, 21
Korvin, Boris V., 88
Kruschka, Henry, 24

La Chapelle, Duval, 16, 22
La Due, William A., 126
Laredo, 67, 69, 74
Laredo, Texas, 67
Larson, Margaret, 112
Larson, Morgan F., 112
Latham, Hubert, 127
Laudenslager, Walter R., 116
Law, Ruth, 44, 45, 47, 100, 107,
  108, 121, 123, 133
Lawrence, Rulif L., 110
Leonard, Herbert, 84
Levick, Edwin, 2
Lodge, Henry Cabot, 26
Loudenslager, Henry C., 26
Lucke, Henry J., 54, 59
Luke Field, 9
Lyons, Thomas B., 119

Mack, Johnny, 16, 26
Mahoney, Daniel E., 83
Manna, William, 68, 113
Margerum, Mahlon R., 13, 14,
  15, 22, 27
McAleenan, Arthur, Jr., 121, 123
McCaffrey, John, 13, 35
McDermott, Joseph, 26, 69, 83,
  115
McPeters, Edward, 83
Mellish, Jay A., 120

Meyer, Edward, 130, 131, 132,
  133, 134
Mickens, Arthur, 84
Mickens, Walter, 84
Moore, Clarence L., 49, 50, 51,
  53, 54, 85
Moore, Thomas, 49
Morford, Leon A., 136
Morgan, Frank, 100
Morgan, Stuart A., 100, 118,
  119, 120, 121
Morris Park, 4
Morris, Helen, 135
Morris, M. Frank, 135, 136
Morris, William, 135
Murgatroyd, Sam, 24

Neely, Henry M., 33
Northam, Nettie, 118

O'Brien, Clarence A., 61
O'Grady, Gerald, 24
Obert, Budd H., Dr., 33
Oettinger, Johanna C., 54
Olsen, Lars, 102, 103, 110
Olson, Louis, 113
Owens, Fred, 16

Pach, Morton V., 113
Parsons, Alonzo R., 16
Parsons, Theodore, 113, 121
patent, x, 28, 41, 50, 51, 52, 53,
  54, 55, 58, 59, 60, 61, 62, 63,
  88, 90, 91, 93, 95, 96, 97,
  118, 131
Patterson, Georgianna, 116
Patterson, Joseph C., 32
Paulhan, Louis, 31
Pepin, Alexander L., 34
Pepin, Emma, 34
Phillips, Dorothy, 103
Pintard, Charles, 41
Pittenger, George W., 16
Poland, Bernard V., 57
Pond, George R., 49
Powers, Harold E., 135

Pratt, Thomas H., Dr., 16
Prinz, Benjamin, 26

Quigley, Timothy, 102

Ralston, James M., 16
Rawlings. L.V., 49
Red Bank Airport, 101, 115
Redden, Charles F., 90
Reid, Arnold, 24
Remond, Edith J., 96
Reyburn, John E., 26
Robinson, Hugh A., 94, 95, 96
Rockafeller, Harry J., 16
Rodriguez, Fausto, 82
Ross, Milan, 16
Ruppert, E. R., 61

Sager, Lawrence., 54
Santos-Dumont, Alberto, 6, 7
Schneider, A. K., 54
Schroeder, Charles J., 138
Schroeder, Edward J., 138
Seidler, Frank, 78
Seligman, Jessie, 96
Sherf, Charles G., Dr., 109
Sickles, Arthur G., 102
Sickles, Lloyd, 113
Simmons, Oliver G., 67, 68, 69, 70, 71, 72, 74, 75
Sloane, John E., 76, 77
Snyder, E. H., 58
Sopwith, Margaret, 73
Sopwith, Thomas, 73, 74
Soule, Orrin G., 137
Steinbach, Walter, 120
Steiner, Harold A., 120
Stevens, George W., 60
Stevens, Russell L., 53
Stevens, Stephen F., 60, 61
Storck, Frank C., 130, 133
Stover, J. Homer, 83
Stuart, Edwin S., Gov., 26
Supp, Henry N., 133

Taylor, Adele G., 1

Thompson, Lewis S., 47
Thurston, David, 53
Thurston, Emma, 54
Tilton, Aaron, 47
Tilyou, George C., 134
Treat, Vernon E., 46, 113, 115, 121, 122, 123
Turkington, Dorothy E., 115
Turner, Frank, 121, 123

Uppercu, Inglis M., 82, 83, 85, 86, 87, 89, 90, 91, 97
Uprichard, Marjorie, 115

Van Brunt, Mattie, 115
Van Keuren, Valentine, 114, 115
Van Wickle, Albert, 41, 42

Wagner, James W., 69
Walker, Wesley, 109
Walling, Daniel, 49
Walling, David A., 48
Walling, Elizabeth, 42, 44, 45
Walling, Eugene, 12
Walling, James, 49
Walling, Margaret Anne, 12
Walling, Peter, 40, 42, 43, 49
Walling, Thomas M., 12, 40, 41, 42, 43, 44, 47, 48, 49
Wanhope, Joshua, 42, 43
Wanhope, Sadie Monro, 42
Warner, James G., 16
Welsh, Alfred L., 74, 75, 76
Wilkins, Wellington, 42
Wilson, Edmund, 23, 24
Wilson, Woodrow, Gov., 70, 71, 72
Winters, Charles D., 121
Wise, Edward, 110
Wittemann, Adolph, 4, 56
Wittemann, Charles R., 4, 5, 56, 118
Wombok, 131
Wombok Second, 134
Wood, Cornelia, 129

Wood, Frank S., 129
Wood, Stephen W., 127, 129, 130
Woolley, Thomas, 24
Wooster, Julian, 54
Worden, Albert W., 104
Wright, Orville, ix, x, 14, 35, 66
Wright, Wilbur, ix, x, 1, 12, 14, 15, 16, 19, 21, 23, 27, 28, 29, 30, 31, 32, 35, 66

Wyckoff, William, 113

Yeager, M. V., 58

Zacharias, Charles R., 16
Zimmerman, Paul G., 91, 92, 93, 94

www.ingramcontent.com/pod-product-compliance
Lightning Source LLC
Chambersburg PA
CBHW050822160426
43192CB00010B/1855